Vegetable Crops

Vegetable Crops

Cole Smith

Larsen & Keller
www.larsen-keller.com

Vegetable Crops
Cole Smith
ISBN: 978-1-64172-117-2 (Hardback)

🖳 Larsen & Keller

Published by Larsen and Keller Education,
5 Penn Plaza,
19th Floor,
New York, NY 10001, USA

Cataloging-in-Publication Data

Vegetable crops / Cole Smith.
     p. cm.
Includes bibliographical references and index.
ISBN 978-1-64172-117-2
1. Vegetables. 2. Food crops. 3. Vegetable gardening. I. Smith, Cole.
SB321 .V44 2019
641.303--dc23

For more information regarding Larsen and Keller Education and its products, please visit the publisher's website www.larsen-keller.com

Table of Contents

Preface

Vegetables are a staple for a healthy diet. They are low in fat and calories, but supply essential dietary fibers, vitamins and minerals. Vegetables can be perennials, biennials or annual. The soil type, climate, rainfall patterns, temperature, length of day, etc. have an influence on vegetable cultivation. The growing season for vegetables can be lengthened by using plastic mulch, fleece, cloches, greenhouses and polytunnels. The tools used in cultivation in domestic and commercial establishments are varied. Spade, hoe, fork, drill, transplanter, irrigation equipment, plough, etc. are some of such tools. Post-harvest care is important for ensuring long storage durability of vegetables. Most vegetables are short lived. Cold storage, shade storing and storage in high-humidity areas are commonly practiced to appropriately store vegetables. These ensure that their nutritional value remains intact. Irradiation of vegetables using ionizing radiation is used for preservation from physical deterioration, microbial infection and insect damage. This book provides comprehensive insights into the production and cultivation of vegetable crops. It picks up aspects of cultivation, harvesting, storage and preservation of vegetable crops and explains their need and contribution in the context of a growing economy. This book will serve as a reference to a broad spectrum of readers interested in this domain.

A foreword of all chapters of the book is provided below:

Chapter 1, Vegetables are grown for consumption by humans. Many different types of vegetables are grown seasonally and annually. These can be root vegetables, stem vegetables, leafy vegetables, flower vegetables, fruit vegetables, seed vegetables, etc. This is an introductory chapter which will discuss briefly about vegetable crops and farming and discusses these diverse types of vegetables and their forming techniques; **Chapter 2**, Leaf vegetables or vegetable greens are plant leaves, petioles and shoots that are eaten as a vegetable. This chapter has been carefully written to provide an easy understanding of the common forms of leaf vegetables, such as spinach, sissoo spinach, celery, Chinese cabbage, etc.; **Chapter 3**, Root vegetables are underground parts of a plant that are consumed by humans. These can be of various types, such as bulb, modified plant stem, root-like stem and true roots. The topics elaborated in this chapter address these different types of root vegetables; **Chapter 4**, Fruit vegetables are seed-bearing parts of plants that are used as vegetables in different cuisines. Some of the common types of fruit vegetables, such as tomatoes, eggplants, cucumbers, zucchinis, pumpkins, etc. have been carefully analyzed in this chapter; **Chapter 5**, Stem vegetables are stems of plants that are eaten as vegetables. Plant inflorescences, such as flowers and flower buds and their stems and leaves that are eaten as vegetables are called inflorescence vegetables. This chapter closely examines the key concepts of stem vegetables and inflorescence vegetables, and their various types; **Chapter 6**, The scientific study of diseases in plants that is caused by pathogens or physiological factors is under the science of plant pathology. Bacteria, viruses, fungi, protozoa, etc. can cause diseases in plants. Black dot, beet vascular necrosis, Tobamovirus, Alternaria solani, etc. are some of the common diseases that affect vegetable production. These diseases along with others have been extensively covered in this chapter; **Chapter 7**, Vegetables are preserved to extend their availability for consumption. Deterioration in quality occurs as a result of the action of microorganisms or due to naturally occurring enzymes. Some of the different vegetable preservation methods are freezing, canning, pickling and fermenting of vegetables which have been carefully analyzed in this chapter.

At the end, I would like to thank all the people associated with this book devoting their precious time and providing their valuable contributions to this book. I would also like to express my gratitude to my fellow colleagues who encouraged me throughout the process.

Cole Smith

Vegetable Crops and Farming

Vegetables are grown for consumption by humans. Many different types of vegetables are grown seasonally and annually. These can be root vegetables, stem vegetables, leafy vegetables, flower vegetables, fruit vegetables, seed vegetables, etc. This is an introductory chapter which will discuss briefly about vegetable crops and farming and discusses these diverse types of vegetables and their forming techniques.

Vegetable Farming

Vegetable farming refers to the growing of vegetable crops, primarily for use as human food.

The term vegetable in its broadest sense refers to any kind of plant life or plant product; in the narrower sense however, it refers to the fresh, edible portion of a herbaceous plant consumed in either raw or cooked form. The edible portion may be a root, such as rutabaga, beet, carrot, and sweet potato; a tuber or storage stem, such as potato and taro; the stem, as in asparagus and kohlrabi; a bud, such as brussels sprouts; a bulb, such as onion and garlic; a petiole or leafstalk, such as celery and rhubarb; a leaf, such as cabbage, lettuce, parsley, spinach, and chive; an immature flower, such as cauliflower, broccoli, and artichoke; a seed, such as pea and lima bean; the immature fruit, such as eggplant, cucumber, and sweet corn; or the mature fruit, such as tomato and pepper.

The popular distinction between vegetable and fruit is difficult to uphold. In general, those plants or plant parts that are usually consumed with the main course of a meal are popularly regarded as vegetables, while those mainly used as desserts are considered fruits. This distinction is applied in this article. Thus, cucumber and tomato, botanically fruits, since they are the portion of the plant containing seeds, are commonly regarded as vegetables.

Types of Production

Vegetable production operations range from small patches of crops, producing a few vegetables for family use or marketing, to the great, highly organized and mechanized farms common in the most technologically advanced countries.

In technologically developed countries the three main types of vegetable farming are based on production of vegetables for the fresh market, for canning, freezing, dehydration, and pickling, and to obtain seeds for planting.

Production for the Fresh Market

This type of vegetable farming is normally divided into home gardening, market gardening, truck farming, and vegetable forcing.

Home gardening provides vegetables exclusively for family use. About one-fourth of an acre of land is required to supply a family of six. The most suitable vegetables are those producing a large yield per unit of area. Bean, cabbage, carrot, leek, lettuce, onion, parsley, pea, pepper, radish, spinach, and tomato are desirable home garden crops.

Market gardening produces assorted vegetables for a local market. The development of good roads and of motor trucks has rapidly extended available markets; the market gardener, no longer forced to confine his operations to his local market, often is able to specialize in the production of a few, rather than an assortment, of vegetables; a transformation that provides the basis for a distinction between market and truck gardening in the mid-20th century. Truck gardens produce specific vegetables in relatively large quantities for distant markets.

In the method known as forcing, vegetables are produced out of their normal season of outdoor production under forcing structures that admit light and induce favourable environmental conditions for plant growth. Greenhouses, cold frames, and hotbeds are common structures used. Hydroponics, sometimes called soilless culture, allows the grower to practice automatic watering and fertilizing, thus reducing the cost of labour. To successfully compete with other fresh market producers, greenhouse vegetable growers must either produce crops when the outdoor supply is limited or produce quality products commanding premium prices.

Production for Processing

Processed vegetables include canned, frozen, dehydrated, and pickled products. The cost of production per unit area of land and per ton is usually less for processing crops than for the same crops grown for market because raw material appearance is not a major quality factor in processing. This difference allows lower land value, less hand labour, and lower handling cost. Although many kinds of vegetables can be processed, there are marked varietal differences within each species in adaptability to a given method.

Specifications for vegetables for canning and freezing usually include small size, high quality, and uniformity. For many kinds of vegetables, a series of varieties having different dates of maturity is required to ensure a constant supply of raw material, thus enabling the factory to operate with an even flow of input over a long period. Acceptable processed vegetables should have a taste, odour, and appearance comparable with the fresh product, retain nutritive values, and have good storage stability.

Vegetables Raised for Seed Production

This type of vegetable farming requires special skills and techniques. The crop is not ready for harvest when the edible portion of the plant reaches the stage of maturity; it must be carried through further stages of growth. Production under isolated conditions ensures the purity of seed yield. Special techniques are applied during the stage of flowering and seed development and also in harvesting and threshing the seeds.

Production Factors and Techniques

Profitable vegetable farming requires attention to all production operations, including insect, disease, and weed control and efficient marketing. The kind of vegetable grown is mainly determined by consumer demands, which can be defined in terms of variety, size, tenderness, flavour, freshness, and type of pack. Effective management involves the adoption of techniques resulting in a steady flow of the desired amount of produce over the whole of the natural growing season of the crop. Many vegetables can be grown throughout the year in some climates, although yield per acre for a given kind of vegetable varies according to the growing season and region where the crop is produced.

Climate

Climate involves the temperature, moisture, daylight, and wind conditions of a specific region. Climatic factors strongly affect all stages and processes of plant growth.

Temperature

Temperature requirements are based on the minimum, optimum, and maximum temperatures during both day and night throughout the period of plant growth. Requirements vary according to the type and variety of the specific crop. Based on their optimum temperature ranges, vegetables may be classed as cool-season or warm-season types. Cool-season vegetables thrive in areas where the mean daily temperature does not rise above 70° F (21°C). This group includes the artichoke, beet, broccoli, brussels sprouts, cabbage, carrot, cauliflower, celery, garlic, leek, lettuce, onion, parsley, pea, potato, radish, spinach, and turnip. Warm-season vegetables, requiring mean daily temperature of 70° F or above, are intolerant of frost. These include the bean, cucumber, eggplant, lima bean, okra, muskmelon, pepper, squash, sweet corn (maize), sweet potato, tomato, and watermelon.

Premature seeding, or bolting, is an undesirable condition that is sometimes seen in fields of cabbage, celery, lettuce, onion, and spinach. The condition occurs when the plant goes into the seeding stage before the edible portion reaches a marketable size. Bolting is attributed to either extremely low or high temperature conditions in combination with inherited traits. Specific vegetable strains or varieties may exhibit significant differences in their tendency to bolt.

Young cabbage or onion plants of relatively large size may bolt upon exposure to low temperatures near 50° to 55° F (10° to 13°C). At high temperatures of 70° to 80° F (21° to 27°C) lettuce plants do not form heads and will show premature seeding. The fruit sets of tomatoes are adversely affected by relatively low and relatively high temperatures. Tomato breeders, however, have developed several new varieties, some setting fruits at a temperature as low as 40° F (4°C) and others at a temperature as high as 90° F (32°C).

Moisture

The amount and annual distribution of rainfall in a region, especially during certain periods of development, affects local crops. Irrigation may be required to compensate for insufficient rainfall. For optimum growth and development, plants require soil that supplies water as well as nutrients dissolved in water. Root growth determines the extent of a plant's ability to absorb water and nutrients, and in dry soil root growth is greatly retarded. Extremely wet soil also retards root growth by restricting aeration. Atmospheric humidity, the moisture content of the air, also contributes moisture. Certain seacoast areas characterized by high humidity are considered especially adapted to the production of such crops as the artichoke and lima bean. High humidity, however, also creates conditions favourable for the development of certain plant diseases.

Daylight

Light is the source of energy for plants. The response of plants to light is dependent upon light intensity, quality, and daily duration, or photoperiod. The seasonal variation in day length affects the growth and flowering of certain vegetable crops. Continuation of vegetative growth, rather than early flower formation, is desirable in such crops as spinach and lettuce. When planted very late in the spring, these crops tend to produce flowers and seeds during the long days of summer before they attain sufficient vegetative growth to produce maximum yields. The minimum photoperiod required for formation of bulbs in garlic and onion plants differs among varieties, and local day length is a determining factor in the selection of varieties.

Each of the climatic factors affects plant growth, and can be a limiting factor in plant development. Unless each factor is of optimum quantity or quality, plants do not achieve maximum growth. In addition to the importance of individual climatic factors, the interrelationship of all environmental factors affects growth.

Certain combinations may exert specific effects. Lettuce usually forms a seedstalk during the long days of summer, but the appearance of flowers may be delayed, or even prevented, by relatively low temperature. An unfavourable temperature combined with unfavourable moisture conditions may cause the dropping of the buds, flowers, and small fruits of the pepper, reducing the crop yield. Desirable areas for muskmelon production are characterized by low humidity combined with high temperature. In the production of seeds of many kinds of vegetables, absence of rain, or relatively light rainfall, and low humidity during ripening, harvesting, and curing of the seeds are very important.

Site

The choice of a site involves such factors as soil and climatic region. In addition, with the continued trend toward specialization and mechanization, relatively large areas are required for commercial production, and adequate water supply and transportation facilities are essential. Topography—that is, the surface of the soil and its relation to other areas—influences efficiency of operation. In modern mechanized farming, large, relatively level fields allow for lower operating costs. Power equipment may be used to modify topography, but the cost of such land renovation may be prohibitive. The amount of slope influences the type of culture possible. Fields with a moderate slope should be contoured, a process that may involve added expense for the building of terraces and diversion ditches. The direction of a slope may influence the maturation time of a crop or may result in drought, winter injury, or wind damage. A level site is generally most desirable, although a slight slope may assist drainage. Exposed sites are not suitable for vegetable farming because of the risk of damage to plants by strong winds.

The soil stores mineral nutrients and water used by plants, as well as housing their roots. There are two general kinds of soils—mineral and the organic type called muck or peat. Mineral soils include sandy, loamy, and clayey types. Sandy and loamy soils are usually preferred for vegetable production. Soil reaction and degree of fertility can be determined by chemical analysis. The reaction of the soil determines to a great extent the availability of most plant nutrients. The degree of acid, alkaline, or neutral reaction of a soil is expressed as the pH, with a pH of 7 being neutral, points below 7 being acid, and those above 7 being alkaline. The optimum pH range for plant growth varies from one crop to another. A soil can be made more acid, or less alkaline, by applying an acid-producing chemical fertilizer such as ammonium sulfate.

The inherent fertility of soils affects production quantity, and a sound fertility program is required to maintain productivity. The ability of a soil to support plant life and produce abundant harvests is dependent on the immediately available nutrients in the soil and on the rate of release of additional nutrients that are present but not available to plants. The rate of release of these additional nutrients is affected by such factors as microbial action, soil temperature, soil moisture, and aeration. Depletion of soil fertility may occur as a result of crop removal, erosion, leaching, and volatilization, or evaporation, of nutrients.

Soil Preparation and Management

Soil preparation for vegetable growing involves many of the usual operations required for other crops. Good drainage is especially important for early vegetables because wet soil retards development. Sands are valuable in growing early vegetables because they are more readily drained than the heavier soils. Soil drainage accomplished by means of ditches or tiles is more desirable than the drainage obtained by planting crops on ridges because the former not only removes the excess water but also allows air to enter the soil. Air is essential to the growth of crop plants and to certain beneficial soil organisms making nutrients available to the plants.

When crops are grown in succession, soil rarely needs to be plowed more than once each year. Plowing incorporates sod, green-manure crops, and crop residues in the soil; destroys weeds and insects; and improves soil texture and aeration. Soils for vegetables should be fairly deep. A depth of six to eight inches (15 to 20 centimetres) is sufficient in most soils.

Soil management involves the exercise of human judgment in the application of available knowledge of crop production, soil conservation, and economics. Management should be directed toward producing the desired crops with a minimum of labour. Control of soil erosion, maintenance of soil organic matter, the adoption of crop rotation, and clean culture are considered important soil-management practices.

Soil erosion, caused by water and wind, is a problem in many vegetable-growing regions because the topsoil is usually the richest in fertility and organic matter. Soil erosion by water can be controlled by various methods. Terracing divides the land into separate drainage areas, with each area having its own waterway above the terrace. The terrace holds the water on the land, allowing it to soak into the soil and reducing or preventing gullying. In the contouring system, crops are planted in rows at the same level across the field. Cultivation proceeds along the rows rather than up and down the hill. Strip cropping consists of growing crops in narrow strips across a slope, usually on the contour. Soil erosion by wind can be controlled by the use of windbreaks of various kinds, by keeping the soil well supplied with humus, and by growing cover crops to hold the soil when the land is not occupied by other crops.

Maintenance of the organic-matter content of the soil is essential. Organic matter is a source of plant nutrients and is valuable for its effect on certain properties of the soil. Loss of organic matter is the result of the action of micro-organisms that gradually decompose it to carbon dioxide. The addition of manures and the growing of soil-improving crops are efficient means of supplying soil organic matter. Soil-improving crops are grown solely for the purpose of preparing the soil for the growth of succeeding crops. Green-manure crops, grown especially for soil improvement, are turned under while still green and usually are grown during the same season of the year as the vegetable crops. Cover crops, raised for both soil protection and improvement, are only grown during seasons when vegetable crops do not occupy the land. When a soil-improving crop is turned under, the various nutrients that have contributed to the growth of the crop are returned to the soil, adding a quantity of organic matter. Both legumes, those plants such as peas and beans having fruits and seeds formed in pods, and nonlegumes are effective soil-improving crops. The legumes, however, are more valuable, because they contribute nitrogen as well as humus. The rate of decomposition of plant material depends on the kind of crop, its stage of growth, and soil temperature and moisture. The more succulent the material is at the time it is turned under, the more quickly it decomposes. Because dry material decomposes more slowly than green material, it is desirable to turn under soil-improving crops before they are mature, unless considerable time is to elapse between the plowing and the planting of the succeeding crop. Plant material decomposes most rapidly when the soil is warm and well supplied with moisture. If soil is dry when a soil-improving crop is turned under, little or no decomposition will occur until rain or irrigation supplies the necessary moisture.

The chief benefits derived from crop rotation are the control of disease and insects and the better use of the resources of the soil. Rotation is a systematic arrangement for the growing of different crops in a more or less regular sequence on the same land. It differs from succession cropping in that rotation cropping covers a period of two, three, or more years, while in succession cropping two or more crops are grown on the same land in one year. In many regions vegetable crops are grown in rotation with other farm crops. Most vegetables grown as annual crops fit into a four-or five-year rotation plan. The system of intercropping, or companion cropping, involves the growing of two or more kinds of vegetables on the same land in the same growing season. One of the vegetables must be a small-growing and quick-maturing crop; the other must be larger and late maturing.

In the practice of clean culture, commonly followed in vegetable growing, the soil is kept free of all competing plants through frequent cultivation and the use of protective coverings, or mulches, and weed killers. In a clean vegetable field the possibility of attack by insects and disease-incitant organisms, for which plant weeds serve as hosts, is reduced.

Propagation

Propagation of crop plants, involving the formation and development of new individuals in the establishment of new plantings, is usually accomplished by the use of either seeds or the vegetative parts of plants. The first type, known as sexual propagation, is used for asparagus, bean, broccoli, cabbage, carrot, cauliflower, celery, cucumber, eggplant, leek, lettuce, lima bean, okra, onion, muskmelon, parsley, pea, pepper, pumpkin, radish, spinach, sweet corn, squash, tomato, turnip, and watermelon. The second type, asexual propagation, is used for the artichoke, garlic, girasole, potato, rhubarb, and sweet potato.

Although seed cost is a small portion of the total cost of crop production, seed quality strongly affects crop success or failure. Good seed should be accurately labelled, clean, graded to size, viable, and free of diseases and insects. The reliability of the seed house is an important factor in obtaining good-quality seed. Viability, or ability to grow, and longevity, the period of viability, are characteristics of seeds of any vegetable kind. In cool, dry storage conditions, those vegetable seeds having comparatively short longevity of one to two years are okra, onion, parsley, and sweet corn. Seeds having three-year longevity are those of the asparagus, bean, carrot, leek, and pea; four-year longevity is characteristic of the beet, chard, pepper, pumpkin, and tomato seeds; longevity of five years characterizes the seeds of broccoli, cabbage, cauliflower, celery, cucumber, eggplant, lettuce, muskmelon, radish, spinach, squash, turnip, and watermelon. The dry seeds of all vegetables, when packed under vacuum in hermetically sealed cans, should remain viable for a longer period than seeds stored under less protective conditions.

Crops grown from hybrid seeds yield vegetables of high quantity and quality. The hybrid-seed industry is based on the production of new seed each year from the controlled pollination of selected parents found to produce the desired combination of characters in the progeny. In the early 1980s the number of F_1 hybrids was increasing in Japan, the United States, and other technically advanced countries. The number of F_1 hybrids varied with the kind of vegetable, but none had yet been introduced for the bean, celery, lettuce, okra, parsley, or pea.

Planting

Most vegetable crops are planted in the field where they are to grow to maturity. A few kinds are commonly started in a seedbed, established in the greenhouse or in the open, and transplanted as seedlings. Asparagus seeds are planted in a seedbed to produce crowns used for field setting. Some vegetables can be either directly seeded in the field or grown from transplants. These include broccoli, cabbage, cauliflower, celery, eggplant, leek, lettuce, onion, pepper, and tomato. The time and method of planting seeds and plants of a particular vegetable influence the success or failure of the crop. Important factors include the depth of planting, the rate of planting, and the spacing both between rows and between plants within a row.

Factors to be considered in determining the time of planting include soil and weather conditions, kind of crop, and desired harvest time. When more than one planting of a crop is made, the second and later plantings should be timed to provide a continuous harvest for the period desired. The soil temperature required for germination of the planted seed varies markedly with the various kinds of vegetables. Vegetables that will not germinate at a temperature below 60° F (16°C) include the bean, cucumber, eggplant, lima bean, muskmelon, okra, pepper, pumpkin, squash, and watermelon. Temperatures higher than 90° F (32°C) are not favourable for the germination of seeds of celery, lettuce, lima bean, parsley, pea, and spinach.

The quantity of seeds planted, or rate of planting, is mainly determined by the characteristics of the vegetable plant. The size of seeds affects the number of plants raised in a given area. Watermelon varieties, for example, differ in seed size expressed as weight. The Sugar Baby variety has an average weight of 1.4 ounces (41 grams) for 1,000 seeds; those of Blackstone variety average 4.4 ounces (125 grams). If the two are grown on two separate plots of the same area and 4.4 ounces of seeds of each cultivar are planted, the result would be three times as many of the Sugar Baby plants as the Blackstone type. Seed size and plant-growth pattern of a vegetable are major factors that govern the number of plants raised in a given area. The trend in the early 1980s was to increase plant population for many crops to achieve the greatest yield possible without impairing quality. As plant population increases per unit area, a point is reached at which each plant begins to compete for certain essential growth factors—*e.g.,* nutrients, moisture, and light. When the population is below the level in which competition between plants occurs, increased population will have no effect on individual plant performance, and the yield per unit area will increase in direct proportion to the increment of population. When competition for essential growth factors occurs, however, yield per plant decreases.

Early harvest and economical use of space are the principal objectives of growing vegetable crops from transplants produced in a greenhouse or outdoor seedbed. It is easier to care for young plants of the cabbage, cauliflower, celery, onion, and tomato in small seedbeds than to sow the seeds in the place where the crop is to grow and mature. Land is free longer for another crop, and weeds, insects, diseases, and irrigation are more readily and economically controlled. The production of transplants is often a specialty of growers who sell their produce to other vegetable growers. The seeds may be planted at a rate three to six times that commonly used for a direct-seeded field. The young plants are removed for use as transplants when they reach the desired size and age, approximately 40 to 60 days after seeding.

Care of Crops during Growth

Practices required for a vegetable crop growing in the field include cultivation; irrigation; application of fertilizers; control of weeds, diseases, and insects; protection against frost; and the application of growth regulators if necessary.

Cultivation

Cultivation refers to stirring the soil between rows of vegetable plants. Because weed control is the most important function of cultivation, this work should be performed at the most favourable time for weed killing, when the weeds are breaking through the soil surface. When the plants are grown on ridges, it is necessary to cover the basal plant portion with soil in the case of such vegetables as asparagus, carrot, garlic, leek, onion, potato, sweet corn, and sweet potato.

Irrigation

Vegetable production requires irrigation in arid and semi-arid regions, and irrigation is frequently used as insurance against drought in more humid regions. In areas having intermittent rain for five or six months, with little or none during the remainder of the year, irrigation is essential throughout the dry season and may also be needed between rainfalls in the rainy season. The two types of land irrigation generally suited to vegetables are surface irrigation and sprinkler irrigation. A level site is required for surface irrigation, in which the water is conveyed directly over the field in open ditches at a slow, nonerosive velocity. Where water is scarce, pipelines may be used, eliminating losses caused by seepage and evaporation. The distribution of water is accomplished by various control structures, and the furrow method of surface irrigation is frequently employed because most vegetable crops are grown in rows. Sprinkler irrigation conveys water through pipes for distribution under pressure as simulated rain.

Irrigation requirements are determined by both soil and plant factors. Soil factors include texture, structure, water-holding capacity, fertility, salinity, aeration, drainage, and temperature. Plant factors include type of vegetable, density and depth of the root system, stage of growth, drought tolerance, and plant population.

Fertilizer Application

Soil fertility is the capacity of the soil to supply the nutrients necessary for good crop production, and fertilizing is the addition of nutrients to the soil. Chemical fertilizers may be used to supply the needed nitrogen, phosphorus, and potassium. Chemical tests of soil, plant, or both are used to determine fertilizer needs, and the rate of application is usually based on the fertility of the soil, the cropping system employed, the kind of vegetable to be grown, and the financial return that might be expected from the crop. Methods of fertilizer application include scattering and mixing with the soil before planting; application with a drill below the surface of the soil at the time of planting; row application before or at planting time; and row application during plant growth, also called side-dressing. Plowed down broadcast fertilizers have recently been used in combination with high analysis liquid fertilizers applied at planting or as a side-dressed band. Mechanical planting devices may employ fertilizer attachments to plant the fertilizer in the form of bands near the seed. For most vegetables, the bands are placed from two to three inches (five to 7.5 centimetres) from the seed, either at the same depth or slightly below the seed.

Weed Control

Weeds (plants growing where they are not wanted) reduce crop yield, increase production cost, and may harbour insects and diseases that attack crop plants. Methods employed to control weeds include hand weeding, mechanical cultivation, application of chemicals acting as herbicides, and a combination of mechanical and chemical means. Herbicides, selective chemical weed killers, are absorbed by the plant and induce a toxic reaction. The amount and type of herbicide that can be safely used to protect vegetable crops depends on the tolerance of the specific crops to the chemical. Most herbicides are applied as a spray, and the appropriate time for application is determined by the composition of the herbicide and the kind of vegetable crop to be treated. Preplanting treatments are applied before the crop is planted; preemergence treatments are applied after the crop is planted but before its seedlings emerge from the soil; and postemergence treatments are applied to the growing crop at a definite stage of growth.

Disease and Insect Control

The production of satisfactory crops requires rigorous disease- and insect-control measures. Crop yield may be lowered by disease or insect attack, and when plants are attacked at an early stage of growth the entire crop may be lost. Reduction in the quality of vegetable crops may also be caused by diseases and insects. Grades and standards for market vegetables usually specify strict limits on the amount of disease and insect injury that may be present on vegetables in a designated grade. Vegetables remain vulnerable to insect and disease damage after harvesting, during the marketing and handling processes. When a particular plant pest is identified, the grower can select and apply appropriate control measures. Application of insect control at the times specific insects usually appear or when the first insects are noticed is usually most effective. Effective disease control usually requires preventive procedures.

Diseases are incited by such living organisms as bacteria, fungi, and viruses. Harmful material enters the plant, develops during an incubation period, and finally causes infection, the reaction of the plant to the pathogen, or disease-producing organism. Control is possible during the inoculation and incubation phases, but when the plant reaches the infection stage it is already damaged. Typical plant diseases include mildew, leaf spots, rust, and wilt. Chemical fungicides may be used to control disease, but the use of disease-resistant plant varieties is the most effective means of control.

Vegetable breeders have developed plant varieties resistant to one or more diseases; such varieties are available for the bean, cabbage, cucumber, lettuce, muskmelon, onion, pea, pepper, potato, spinach, tomato, and watermelon.

Insects are usually controlled by the use of chemical insecticides that kill through toxic action. Many insecticides are toxic to harmful insects but do not affect bees, which are valuable for their role in pollination.

Frost Protection

Frost protection may be accomplished by increasing the amount of heat radiated from the soil when frost is likely to occur. Irrigation on the day before a predicted frost provides additional moisture in the soil to increase the amount of heat given off as infrared rays. This extra heat protects the plants from frost injury. A continuous supply of water provided by sprinkler irrigation may also protect plants from frost. As the water freezes on the plant leaves, it loses heat that is absorbed by the plant leaves, maintaining leaf temperature at 32° F (0°C). Because of the sugars and other substances in plant cells, the freezing point of cell sap is somewhat lower than 32° F.

Growth Regulators

It is sometimes desirable to retard or accelerate maturity in vegetable crops. A chemical compound may be applied to prevent sprouting in onion crops. It is applied in the field sufficiently early for absorption by the still-green foliage but late enough to avoid suppressing the bulb yield. Another substance may be used to end the dormancy, or rest period, of newly harvested potato tubers intended for planting. The treated seed potatoes have uniform sprout emergence. The same substance is applied to celery from two to three weeks before harvest to elongate the stalks and increase the yield and is also used to accelerate maturity in artichokes. A chemical compound, applied when adverse weather conditions prevail during the period of fruit setting, has been used to encourage fruit set.

Harvesting

The stage of development of vegetables when harvested affects the quality of the product reaching the consumer. In some vegetables, such as the bean and pea, optimum quality is reached well in advance of full maturity and then deteriorates, although yield continues to increase. Factors determining the harvest date include the genetic constitution of the vegetable variety, the planting date, and environmental conditions during the growing season. Successive harvest dates may be obtained either by planting varieties having different maturity dates or by changing the sequence of planting dates of one particular variety. The successive method is applicable to such crops as broccoli, cabbage, cauliflower, muskmelon, onion, pea, sweet corn, tomato, and watermelon. Certain varieties of the carrot, celery, cucumber, lettuce, parsley, radish, spinach, or summer squash can be sown in succession throughout most of the year in some climates, thus prolonging the harvest period.

Hand harvesting is employed along with various mechanical aids for broccoli, cabbage, cauliflower, muskmelon, and pepper crops. Many vegetables grown for processing and some vegetables destined for the fresh market are mechanically harvested. Harvesting operations may be performed by a single machine in a single step for such vegetable crops as the bean, beet, carrot, lima bean, onion, pea, potato, radish, spinach, sweet corn, sweet potato, and tomato. Designers of harvesting machinery have been working to develop a multiple-picking harvester capable of adjustment for use with more than one crop. Vegetable breeders have been able to produce vegetables with characteristics suitable for machine harvesting, including compact plant growth, uniform development, and concentrated maturity.

Root Vegetables

Root vegetables are plants whose roots are commonly harvested for culinary usage. The term is used casually or unofficially to refer to many types of plants, aside from simply taproot plants and tuberous root plants.

For instance, root vegetables are a massive category of edible plants that include smaller categories like bulbs, tubers, true roots, root-like stems, modified plant stems, corns, rhizomes, and so on.

Popular root vegetables include carrots, beets, potatoes, yams, shallots, onions, garlic, ginseng, and turmeric, etc.

The roots of root vegetables generally act as the plant's storage organ and are often rich in sugars, starches, and other types of carbohydrates.

Root vegetables, which are often considered starchy, are staple foods in regions like Oceania, West Africa, and Central Africa, surpassing cereals and grains in their value.

Root vegetables have been known to have quite a long shelf life, especially if stored in root cellars. In non-tropical latitudes, it is especially important to store these vegetables in root cellars because there are practically no harvests during winter. When growing carrots, gardeners normally choose to go for larger growing bags and pots to ensure proper drainage.

Similarly, beets also need a deep container to thrive in because they have spherical roots. In general, root vegetables should be regularly watered, without over-watering in order to protect the plants from rot and premature withering.

When working with these types of vegetables, it's also important to ensure that the soil is loose enough for the plants to send down their roots.

Stem Vegetables

Plant stems, which are used as vegetables are called, stem vegetables. These can be divided into above the ground edible stems and modified underground stems. Crown, runners, stolons or spurs are stems that are located above the ground. Bulbs, corms, tubers and rhizomes are swollen modified underground stems that serve as storage organs.

Leafy Vegetables

Leafy greens are dark green, shallow-rooted plants that demand plenty of water for proper growth. The term is non-scientific and has risen in popularity as the health benefits of leafy greens have been highly reported for decades.

Common leafy greens include, but are not limited to, herbs, chards, beans, spinach, and various types of lettuces. These plants are extremely easy to grow and do not require a lot of care or soil preparation.

Leafy greens are considered highly suitable for hydroponic gardening due to their quick growth cycle, quick turnaround times, easy marketability, and shallow root zones. When sown in the early stages, leafy greens may be referred to as microgreens.

It is important to know that leafy salad greens tend to germinate quite quickly in humid and warm conditions. Therefore, some greens such as lettuce need plenty of shade to thrive. When working with Asian greens and spinach, it's also important to grow them in non-acidic and fertile soil, preferably in square containers that are properly angled towards the sun.

According to gardeners, spinach and lettuce can dry up and go to seed in extremely hot weather, which can make the leaves turn bitter. In such cases, planters can sporadically water the containers to ensure that the soil remains moist.

While mint is one of the most popular and easiest leafy greens to grow at home, it can become something of a nuisance when grown in a vegetable garden if it's not properly contained.

Climatic Requirements

Leafy vegetables may be cool-season or warm-season crops and can be grown as annuals or as perennials. In addition, some leaf vegetables are adapted to the tropics, while others are adapted to the temperate climates. Leafy vegetables such as, lettuce and spinach are most tolerant of shade; in fact, in locations of hot and bright sun, they may need some shade for protection. These vegetables thrive in areas where the mean temperatures range between 15 and 18°C. They are intolerant to temperatures between 21 and 24°C, and tolerate weak frosts.

Soil Requirements

Loose, fertile, moist, sandy loam soils are best for growing leafy vegetables. Many of leafy vegetables have shallow root systems and cultivation should be done carefully. All leafy vegetables, except lettuce, grow best in soils with a pH of 6, 0 to 6, 8. Lettuce grows best at a pH of 6.5 to 7.0.

Weed Management

Effective weed control for leafy vegetables should include a combination of practices designed to suppress weeds during the entire year. Some of the management practices include crop rotation, cover cropping, cultivation flooding, and mulching. Leafy vegetables can also be mulched with 8 to 10 cm thick layer of herbicide-free grass clippings, or weed-free straw to retain soil moisture and suppress weeds. Care should be taken when the leafy crops are rotated behind crops for which more persistent herbicides are used.

Potential Problems and Management

Aphids can be a problem in leafy vegetables and they can be controlled by hosing the leaves when watering and by natural enemies like ladybird beetles. Cabbage worms and flea beetles may also damage the plant leaves. To prevent such damage, cover the plants with floating row covers when moths and flea beetles are seen flying in the garden. To manage cabbage worm, registered insecticides can also be used but should not be sprayed only over the top of plants because most eggs and younger loopers feed on the underside of leaves, so the underside of the leaf should be sprayed as well. Leafminers can also be a problem in producing leafy vegetables. Plants are often disfigured and damaged by larvae of several species of small flies that live as maggots between the upper and lower surfaces of the leaves. Their feeding causes large, white blotches and winding trails through the interior of the leaves. Eliminating weeds will aid in the management of leafminers during most years but sprays may be needed to prevent damage.

Flower Vegetables

The "Flower vegetables" category includes plants of which the flowers are used as food. Examples of flower vegetables are cauliflower and artichoke.

Total 6 crops found in category Flower vegetables:

1) Agati

2) Artichoke

3) Banana

4) Broccoli

5) Cauliflower

6) Lotus

Fruit Vegetables

Fruit vegetables are those vegetables which are fruit-like in their appearance or properties.

Crop name	Scientific name	Family	Collective name for members of tahe family, other info
		Examples of Fruit Vegetables	
Bottle gourd, upo	*Lagenaria siceraria*	Cucurbitaceae	Cucumber/Gourd family, also called Cucurbits
Breadnut, Seeded breadfruit, camansi	*Artocarpus altilis*	Moraceae	Mulberry family
Charantia, bitter melon, bitter gourd, ampalaya	*Momordica charantia*	Cucurbitaceae	Cucumber/Gourd family, also called Cucurbits; the leaves with young stems are also eaten blanched or as ingredient in many vegetable dishes.
Chayote	*Sechium edule*	Cucurbitaceae	Cucumber/Gourd family, also called Cucurbits; chayote tops consisting of young leaves and stems are also cooked as vegetable.
Cucumber	*Cucumis sativus*	Cucurbitaceae	Cucumber/Gourd family, also called Cucurbits; usually consumed raw but also cooked as an ingredient in some food recipes.
Eggplant, aubergine	*Solanum melongena*	Solanaceae	Nightshade family, also called Solanaceous crops
Jackfruit	*Artocarpus heterophyllus*	Moraceae	Mulberry family; mainly grown for the production of ripe fruit but young fruits are cooked as vegetable.
Luffa, loofah, sponge gourd, patola	*Luffa acutangula*	Cucurbitaceae	Cucumber/Gourd family, also called Cucurbits
Okra, Lady's finger, gumbo	*Abelmoschus esculentus*	Malvaceae	Mallow family

Papaya, pawpaw	*Carica papaya*	Caricaceae	largely grown for the production of ripe fruit but young fruits are commomly cooked alone or as ingredient of many food recipes.
Bell/Sweet Pepper	*Capsicum annuum*, Grossum group	Solanaceae	Nightshade family, also called Solanaceous crops
Hot/Chili Pepper	*Capsicum annuum*, Longum group	Solanaceae	Nightshade family, also called Solanaceous crops
Pumpkin	*Cucurbita pepo*	Cucurbitaceae	Cucumber/Gourd family, also called Cucurbits
Tomato	*Lycopersicon esculentum*	Solanaceae	Nightshade family, also called Solanaceous crops
Squash, kalabasa	*Cucurbita maxima*	Cucurbitaceae	Cucumber/Gourd family, also called Cucurbits
Snap bean, kidney bean	*Phaseolus vulgaris*	Fabaceae/Leguminosae	Bean/Pea family, also called Legumes
String bean, Pole sitao, yardlong bean, sitaw	*Vigna unguiculata*subsp. *sesquipedales*	Fabaceae/Leguminosae	Bean/Pea family, also called Legumes
Sweet corn	*Zea mays*	Poaceae/Gramineae	Grass family; young ears, or "baby corn", and "green corn" are also harvested from flint- and dent-type corn varieties.
Winged bean	*Psophocarpus tetragonolobus*	Fabaceae/Leguminosae	Bean/Pea family, also called Legumes

Seed Vegetables

Seed vegetables are vegetables commonly known as the legume, pea, or bean family.They are a large and economically important family of flowering plants. It includes trees, shrubs, and perennial or annual herbaceous plants, which are easily recognized by their fruit (legume) and their compound, stipulated leaves. Many legumes have characteristic flowers and fruits.

Crop name	Scientific name	Family	Collective name for members of the family, other info
Examples of Flower Vegetables			
Artichoke, globe artichoke	*Cynara scolymus*	Asteraceae/Compositae	Sunflower or Aster family
Broccoli	*Brassica oleracea*, Italica group	Brassicaceae/Cruciferae	Mustard family
Cauliflower	*Brassica oleracea*, Botrytis group	Brassicaceae/Cruciferae	Mustard family
Cooking banana, plantain, cardava	*Musa* sp.	Musaceae	Banana family; largely grown for mature fruits but the male inflorescence is used as an ingredient in the preparation of many vegetable dishes; timely removal of the male inflorescence from developing fruit bunches is an established practice in banana plantations.

West Indian pea, agati, katuray	*Sesbania grandiflora*	Fabaceae/Leguminosae	Bean/Pea family, also called Legumes; newly opened flowers are harvested
Examples of Seed Vegetables			
Cowpea, Black-eyed pea	*Vigna unguiculatasyn. Vigna sinensis*	Fabaceae/Leguminosae	Bean/Pea family, also called Legumes
Fava bean	*Vicia faba*	Fabaceae/Leguminosae	Bean/Pea family
Lima bean, patani	*Phaseolus lunatus*	Fabaceae/Leguminosae	Bean/Pea family
Hyacinth bean, bataw	*Dolichos lablab* syn. *Lablab purpureus*	Fabaceae/Leguminosae	Bean/Pea family
Mung bean, mungbean, mungo	*Vigna radiata*	Fabaceae/Leguminosae	Bean/Pea family
Pea, garden pea, snap pea	*Pisum sativum*	Fabaceae/Leguminosae	Bean/Pea family
Pigeon pea, kadios	*Cajanus cajan*	Fabaceae/Leguminosae	Bean/Pea family

Crop Rotation for Vegetables

Vegetables crops are grouped into families. Crop rotation simply means that related annual vegetables are grown together in their families and their positions moved around the plot once a year (or more).

Need to use Crop Rotation

There are a number of reasons for rotating crops:

- It helps to prevent pests and diseases that live in the soil. For example, two major worries in vegetable growing are clubroot disease in Brassica crops (cabbage type plants) and the nematode known as eelworm in potatoes. If the crops are grown in the same place each year, the chances of these problems occurring are much greater. By moving them around annually and only growing them in the same ground every four years of so, the pest and disease life cycles should be broken.

- It stops the soil becoming drained of nutrients that the same plants would use every year.

- Crops can follow each other that will benefit each other. E.g., bean and pea roots hold lots of nitrogen. If their disease free roots are left in the ground once the crops have been harvested, the Brassica that will follow in the next rotation will reap the rewards by producing lots of leafy greens. Also Brassica like soil that's consolidated so by leaving the legume roots behind and thus causing little disturbance to the soil, the Brassica that follow will root better.

- If vegetable families are grown together, it's likely that the soil for each will need to be treated in the same way and that they will be prone to the same pests and diseases so can be treated together easily.

Methods to do Crop Rotation

Divide your vegetable garden or allotment into sections of equal size (depending on how much of each crop you want to grow), plus an extra section for perennial crops, such as rhubarb and asparagus. Group your crops as below:

- Brassicas: Brussels sprouts, cabbage, cauliflower, kale, kohl-rabi, oriental greens, radish, swede and turnips.

- Legumes: Peas, broad beans (French and runner beans suffer from fewer soil problems and can be grown wherever convenient).

- Onions: Onion, garlic, shallot, leek.

- Potato family: Potato, tomato, (pepper and aubergine suffer from fewer problems and can be grown anywhere in the rotation).

- Roots: Beetroot, carrot, celeriac, celery, Florence fennel, parsley, parsnip and all other root crops, except swedes and turnips, which are brassicas.

Move each section of the plot a step forward every year so that, for example, brassicas follow legumes, onions and roots, legumes, onions and roots follow potatoes and potatoes follow brassicas. Here is a traditional three year rotation plan where potatoes and brassicas are important crops:

- Year one
  - Section one: Potatoes
  - Section two: Legumes, onions and roots
  - Section three: Brassicas

- Year two
  - Section one: Legumes, onions and roots
  - Section two: Brassicas
  - Section three: Potatoes

- Year three
  - Section one: Brassicas
  - Section two: Potatoes
  - Section three: Legumes, onions and roots

For Four-year Rotation

This is a four-year rotation for where potatoes and brassicas are not as important, but more legumes (which take up a lot of space) and onion-type crops are required:

- Year one
  - Section one: Legumes
  - Section two: Brassicas
  - Section three: Potatoes
  - Section four: Onions and roots
- Year two
  - Section one: Brassicas
  - Section two: Potatoes
  - Section three: Onions and roots
  - Section four: Legumes
- Year three
  - Section one: Potatoes
  - Section two: Onions and roots
  - Section three: Legumes
  - Section four: Brassicas
- Year four
  - Section one: Onions and roots
  - Section two: Legumes
  - Section three: Brassicas
  - Section four: Potatoes

Vegetable Harvesting

Harvesting vegetables at the right stage of maturity ensures the best taste and quality. Many vegetables should be picked throughout the summer to maintain plant productivity. The time, frequency, and method of harvesting vary depending on species. Vegetables, such as standard sweet corn, have a very small harvest period. Others, such as many of the root crops, can remain in the garden for several weeks with little effect on their taste. Some vegetables, like summer squash, have to be harvested almost daily. Other plants, such as tomatoes, can be harvested on a weekly basis. Use the table below to determine the optimal time to pick and enjoy your favorite vegetables.

Vegetable Harvest Guide				
Vegetable	*Days to Maturity*	*Size*	*Color*	*Comment*
Beet	50-70	2-3 in diameter	red, varies with cultivar	Up to 1/3 of the beet foliage can be harvested for greens without harming the root.
Broccoli	50-65*	6 to 7 across	blue-green	Harvest before yellow flower buds start to open, side shoots can be harvested after main head is removed.
Cabbage	60-90*	varies with cultivar	green, red	Harvest when heads are large and solid.
Carrot	60-80	3/4 in diameter	orange	Harvest when orange shoulder pushes through the soil.
Cauliflower	55-80*	6 to 8 across	creamy white	Blanch heads when 2-3 across by carefully tying leaves over heads.
Cucumber				
Pickling	55-65	2-4 long	dark green	Harvest plants every 2 to 3 days, leave small piece of stem attached to fruit.
Slicing	55-65	6-8 long	dark green	1 to 2 diameter, harvest plants every 2 to 3 days.
Eggplant	75-90*	varies with cultivar	purple, white, green	Fruit should have shiny finish.
Garlic	90**	2-3	white, reddish purple	Harvest when foliage topples over and dries or just before first frost.
Kohlrabi	55-70	2-3 diameter	green	Store with leaves and roots removed.
Lettuce (leaf)	45-60	4-6 long	green	Harvest outer leaves, hot weather causes bitterness.
Muskmelon Cantaloupe	75-100	5-10 in diameter	yellow-tan between netting	When mature, stem separates easily from melon.
Okra	50-65	3 long	bright green	Harvest frequently to maintain productivity.
Onion	100-120 90-100**	varies with cultivar	white, yellow, red	Harvest when tops fall over and begin to dry.
Parsnip	110-130	8-18" long	white or cream	Can be overwintered in the ground, mulch and dig before new growth starts in spring.
Peas				
Snow (Sugar)	55-85	3 long pods	bright green	Harvest when pods are long and thin, just as the seeds begin to develop.
Snap	55-85	3 long pods	bright green	Pick when seeds are nearly full size.
Garden (Shell)	55-85	3 long pods	bright green	Harvest when peas are full size.

Pepper				
Hot	60-90*	1 to 3 long	red, purple, yellow, green	Use gloves when harvesting.
Sweet	70-90*	2 to 4 in diameter	green, red, yellow, purple, orange	Usually harvested when green, but can be left on plant until red, orange, yellow or purple.
Potato	90-120	varies with cultivar	varies with cultivar	Dig when tops turn brown and die.
Pumpkin	85-120	varies with cultivar	orange	Harvest when uniformly orange, leave 3-4 of stem.
Radish				
Spring	25-40	1/2 to 2 in diameter	red, white, varies	Radishes larger than 2 in diameter are often pithy and unusable.
Winter	45-70	6-12 long	white, varies	Can be left in the ground until frost.
Snap Bean (Green Bean)	50-70	4 to 6 long	green, yellow, purple	Harvest when pods are pencil size in thickness.
Spinach	45-60	6-8 tall	green	Harvest the entire crop when plants begin to show signs of bolting.
Summer squash				
Scallop	50-60	3 to 5 in diameter	yellow, green	Harvest when skin is soft.
Zucchini	50-60	6 to 12 long	green, yellow	Harvest every 2 to 3 days when fruit are 2 in diameter.
Sweet Corn	70-105	5 to 10 , varies with cultivar	yellow, white, bicolor	Mature kernels exude milky sap when punctured.
Sweet Potato	100-125	varies with cultivar	Gold or orange	Harvest just before or after a vine killing frost.
Tomato	70-90*	varies with cultivar	red, orange, yellow	Harvest fully ripe for best flavor.
Turnip	45-70	2-3 in diameter	white, reddish purple	Foliage can be harvested for greens.
Watermelon	80-100	varies with cultivar	light to dark green, striped	Harvest when 'belly' turns from white to creamy yellow.
Winter Squash	85-120	varies with cultivar	varies with cultivar	Rind should be hard and difficult to puncture with fingernail.
* from transplants	** from sets (bulbs)			

Vegetable Storage

The vegetables can be stored, in some specific natural conditions, in fresh state, that is without significant modifications of their initial organoleptic properties. Fresh vegetable storage can be short term; this was briefly covered under temporary storage before processing. Also fresh vegetable storage can be long term during the cold season in some countries and in this case it is an important method for vegetable preservation in the natural state.

In order to assure preservation in long term storage, it is necessary to reduce respiration and transpiration intensity to a minimum possible; this can be achieved by:

1. Maintenance of as low a temperature as possible (down to 0°C),

2. Air relative humidity increased up to 85-95 % and

3. CO_2 percentage in air related to the vegetable species.

Vegetables for storage must conform to following conditions: they must be of one of the autumn or winter type variety; be at edible maturity without going past this stage; be harvested during dry days; be protected from rain, sun heat or wind; be in a sound state and clean from soil; be undamaged.

From the time of harvest and during all the period of their storage vegetables are subject to respiration and transpiration and this is on account of their reserve substances and water content. The more the intensity of these two natural processes are reduced, the longer sound storage time will be and the more losses will be reduced.

For this reason, vegetables have to be handled and transported as soon as possible in the storage conditions (optimal temperature and air relative humidity for the given species). Even in these optimal conditions storage will generate losses in weight which are variable and depend upon the species.

Some optimal storage conditions are shown in table.

Table: Optimal conditions for fresh vegetable storage.

Vegetables	Storage conditions	
	Temperature°C	Relative humidity%
Potatoes	+1...+3	85-90
Carrots	0 ... +1	90-95
Onions	0 ... +1	75-85
Leeks	0 ... +0.5	85-90
Cabbage	-1 ... 0	90-97
Garlic	0 ... +1	85-90
Beets	0 ... +1	90-95

Vegetable Preservation

Vegetable preservation includes all the methods which are used for the purpose of preserving food from getting spoiled.

Here are the primary ways of preserving fruit and vegetables.

Canning

Canning involves placing fruit and vegetables in airtight containers, typically glass jars, and so prevent bacteria getting to them. Canned good can be stored on shelves for years, if required. There are two methods, although one requires a specialist machine so may not be practical or cost-efficient for many people. This is the pressure canning method, which enables you to achieve temperatures above boiling point that foods with low acidity require to effectively neutralise the threat of the botulism bacteria remaining active. It requires a pressure canning machine and is the method used to can most vegetables, as they are low in acid. Fruit, being high in acid, does not have the threat of botulism, so can be canned using a simpler method. Just place your fruit in the jar, top with boiling water, leaving an inch or so of space at the top of the jar, run a spatula around the inside edge to remove any air bubbles, then close with a threaded lid.

Whichever method you use, the jars must be sterilized before being filled. You can do this either by using sterilizing tablets such as those used for babies' bottles, or by placing the jars in an oven on a low temperature for half and hour or so in order to kill all the bacteria.

Salting

One of the oldest methods of preserving food, salting can be used for meat and fish, as well as sliced vegetables. There are two methods. The first uses a low salt to vegetable ratio (between two and five percent salt per weight of vegetables). This level of salting promotes the growth of the lactic acid bacteria, which in turn inhibits the growth of other bacterial forms that could spoil the food. It also serves to slightly pickle the vegetables. The second method uses a higher percentage of salt (between twenty and twenty-five percent), preserving the freshness of the produce but adding a salty flavour when used, even after the salt has been washed off. Whichever method of salting you use, you need to store the produce in the refrigerator.

Drying

Drying dehydrates the fruit or vegetables, removing all the water along with the bacteria, yeasts and mold that live in the moisture. Besides altering the texture of the food, drying also modifies the taste, typically concentrating it. Dried food has the added benefit of being safe to store as is on your pantry shelf – you don't need special packaging to keep it in or to keep it in the refrigerator. In some countries solar drying of food is a part of life, and if you live in an area that receives high levels of consistent sunshine, you may be able to dry food that way. More likely however, is drying in an oven. The technique requires low temperature and good air circulation so use the lowest setting and prop the oven door open – this allows the air that the moisture has evaporated into to escape.

Freezing

Freezing fruit and vegetables soon after they are picked serves to 'lock in' the flavour and freshness of the produce. Freezing and then thawing a vegetable or fruit is the preserving method that will have an end product that most closely resembles the taste of fresh food. You effectively place the food in suspended animation in whatever condition it is in when you freeze it, so always freeze ripe produce, and avoid spoiled specimens. You can freeze the produce in wax-coated cardboard containers, in plastic boxes or jars made with very thick glass. It is recommended that you blanch vegetables you are going to freeze in boiling water for a minute or so beforehand – this limits the activity of enzymes that may spoil the produce if stored over a long time. You need a temperature below freezing point for effective long-term storage, so use the freezer compartment in your refrigerator for food that you will use within a month, as temperatures in these rarely get down to the requisite zero degrees. When thawing food, leave at room temperature until completely thawed, rather than trying to thaw in the oven.

References

- Vegetable-farming: britannica.com, Retrieved 29 July 2018

- Flower-vegetables: world-crops.com, Retrieved 09 May 2018

- Fruit-vegetables: cropsreview.com, Retrieved 18 April 2018

- Crop-rotation-what-does-it-mean: greensideup.ie, Retrieved 30 June 2018

- Vegguide: hortnews.extension.iastate.edu, Retrieved 15 July 2018

- Four-ways-preserve-fruit-vegetables: regenerative.com, Retrieved 29 March 2018

2

Leaf Vegetables

Leaf vegetables or vegetable greens are plant leaves, petioles and shoots that are eaten as a vegetable. This chapter has been carefully written to provide an easy understanding of the common forms of leaf vegetables, such as spinach, sissoo spinach, celery, Chinese cabbage, etc.

Leaf vegetables in super market

Leaf vegetables, also called potherbs, greens, vegetable greens, leafy greens or salad greens, are plant leaves eaten as a vegetable, sometimes accompanied by tender petioles and shoots. Although they come from a very wide variety of plants, most share a great deal with other leaf vegetables in nutrition and cooking methods.

Nearly one thousand species of plants with edible leaves are known. Leaf vegetables most often come from short-lived herbaceous plants such as lettuce and spinach. Woody plants whose leaves can be eaten as leaf vegetables include Adansonia, Aralia, Moringa, Morus, and Toona species.

The leaves of many fodder crops are also edible by humans, but usually only eaten under famine conditions. Examples include alfalfa, clover, and most grasses, including wheat and barley. These plants are often much more prolific than more traditional leaf vegetables, but exploitation of their rich nutrition is difficult, primarily because of their high fiber content. This obstacle can be overcome by further processing such as drying and grinding into powder or pulping and pressing for juice.

Leaf vegetables contain many typical plant nutrients, but since they are photosynthetic tissues, their vitamin K levels in relation to those of other fruits and vegetables, as well as other types of foods, is particularly notable. The reason is that phylloquinone, the most common form of the vitamin, is directly involved in photosynthesis. This causes leaf vegetables to be the primary food class that interacts significantly with the anticoagulant pharmaceutical warfarin.

During the first half of the 20th century, it was common for greengrocers to carry small bunches of herbs tied with a string to small green and red peppers, these bundles were called "potherbs."

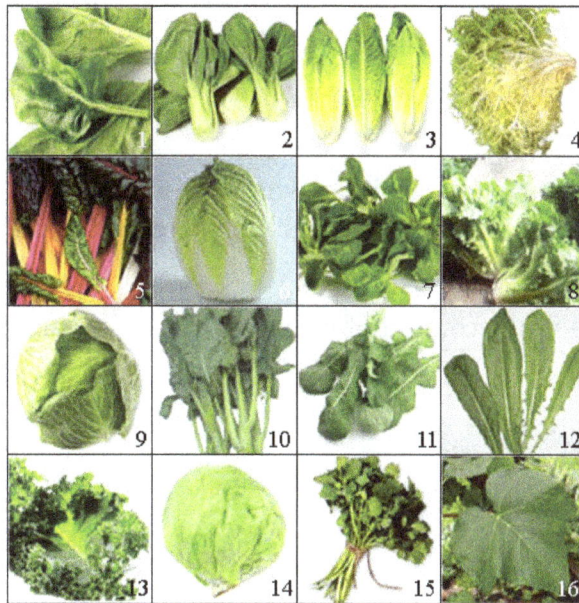

Some kinds of leaf vegetables

Examples of Healthy Leafy Green Vegetables

Kale

Kale is considered one of the most nutrient-dense vegetables on the planet due to its many vitamins, minerals and antioxidants.

For example, one cup (67 grams) of raw kale packs 684% of the Daily Value (DV) for vitamin K, 206% of the DV for vitamin A and 134% of the DV for vitamin C.

It also contains antioxidants such as lutein, carotenoids and beta-carotene, which prevent diseases caused by oxidative stress.

To benefit most from all that kale has to offer, it's best consumed raw since cooking can reduce its nutrient profile.

Microgreens

Microgreens are immature greens produced from the seeds of vegetables and herbs. They typically measure 1–3 inches (2.5–7.5 cm).

Since the 1980s, they have often been used as a garnish or decoration, but they have many more uses.

Despite their small size, they're full of color, flavor and nutrients. In fact, one study found that microgreens contain up to 40 times more nutrients compared to their mature counterparts. Some of these nutrients include vitamins C, E and K.

Microgreens can be grown in the comfort of your own home all year round, making them easily available.

Broccoli

Broccoli is part of the cabbage family.

It has a large flower head as well as a stem, making it similar in structure to cauliflower.

This vegetable is rich in nutrients, with a single cup (91 grams) of raw broccoli packing 135% and 116% of the DVs for vitamins C and K respectively. It's also a great source of fiber, calcium, folate and phosphorus.

Of all vegetables in the cabbage family, broccoli is richest in the plant compound sulforaphane, which may improve your bacterial gut flora and decrease your risk of cancer and heart disease.

What's more, sulforaphane may even reduce symptoms of autism.

One randomized double-blind study in 26 young people with autism observed a positive effect on behavioral symptoms after consuming sulforaphane supplements from broccoli sprouts.

Collard Greens

Collard greens are loose leaf greens, related to kale and spring greens. They have thick leaves that taste slightly bitter.

They're similar in texture to kale and cabbage. In fact, their name comes from the word "colewort," meaning "the wild cabbage plant."

Collard greens are a good source of calcium and the vitamins A, B9 (folate) and C. They're also one of the best sources of vitamin K when it comes to leafy greens. In fact, one cup (190 grams) of cooked collard greens packs 1,045% of the DV for vitamin K.

Vitamin K is known for its role in blood clotting. In addition, more research is being done regarding its ability to improve bone health.

One study in 72,327 women aged 38–63 found that those with vitamin K intakes below 109 mcg per day had a significantly increased risk of hip fractures, suggesting a link between this vitamin and bone health.

Spinach

Spinach is a popular leafy green vegetable and is easily incorporated into a variety of dishes, including soups, sauces, smoothies and salads.

Its nutrient profile is impressive with one cup (30 grams) of raw spinach providing 181% of the DV for vitamin K, 56% of the DV for vitamin A and 13% of the DV for manganese.

It's also packed with folate, which plays a key role in red blood cell production and the prevention of neural tube defects in pregnancy.

One study on the neural tube defect spina bifida found that one of the most preventable risk factors for this condition was a low intake of folate during the first trimester of pregnancy.

Along with taking a prenatal vitamin, eating spinach is a great way to increase your folate intake during pregnancy.

Cabbage

Cabbage is formed of clusters of thick leaves that come in green, white and purple colors.

It belongs to the *Brassica* family, along with Brussels sprouts, kale and broccoli.

Vegetables in this plant family contain glucosinolate, which gives them a bitter flavor.

Animal studies have found that foods that contain this substance may have cancer-protective properties, especially against lung and esophageal cancer.

Another benefit of cabbage is that it can be fermented and turned into sauerkraut, which provides numerous health benefits, such as improving your digestion and supporting your immune system. It may even aid weight loss.

Beet Greens

Since the Middle Ages, beets have been claimed to be beneficial for health.

Indeed, they have an impressive nutrient profile, but while beets are commonly used in dishes, the leaves are often ignored.

This is unfortunate, considering that they're edible and rich in potassium, calcium, riboflavin, fiber and vitamins A and K. Just one cup (144 grams) of cooked beet greens contains 220% of the DV for vitamin A, 37% of the DV for potassium and 17% of the DV for fiber.

They also contain the antioxidants beta-carotene and lutein, which have shown to prevent eye disorders such as muscular degeneration and cataracts.

Beet greens can be added to salads, soups or sauteed and eaten as a side dish.

Watercress

Watercress is an aquatic plant from the *Brassicaceae* family and thus similar to arugula and other mustard greens.

It's known for its healing properties and has been used in medicine for centuries.

Studies have found watercress extract to be beneficial in targeting cancer stem cells and impairing cancer cell reproduction and invasion.

Due to its bitter and slightly spicy flavor, watercress makes a great addition to neutrally flavored foods.

Romaine Lettuce

Romaine lettuce is a common leafy vegetable with sturdy, dark leaves with a firm center rib.

It has a crunchy texture and is a popular lettuce, particularly in Caesar salads.

It's a good source of vitamins A and K, with one cup (47 grams) providing 82% and 60% of the DVs for these vitamins respectively.

What's more, research has found that water intake from fluids, vegetables and fruits plays an important role in weight loss.

Therefore, with only 8 calories and 45 grams of water in a single cup, romaine lettuce may be a great addition to a healthy diet if you're trying to lose weight.

Swiss Chard

Swiss chard has dark-green leaves with a thick stalk that is red, white, yellow or green. It's often used in Mediterranean cooking and belongs to the same family as beets and spinach.

It has an earthy taste and is rich in minerals and vitamins, such as potassium, manganese and the vitamins A, C and K.

Swiss chard also contains a unique flavonoid called syringic acid — a compound that may be beneficial for lowering blood sugar levels.

In two small studies in rats with diabetes, oral administration of syringic acid for 30 days improved blood sugar levels.

However, it's important to note that these were minor animal studies and that human research supporting the claim that syringic acid may aid blood sugar control is lacking.

While many people typically throw away the stems of the Swiss chard plant, they're crunchy and highly nutritious.

Next time, try adding all parts of the Swiss chard plant to dishes such as soups, tacos or casseroles.

Arugula

Arugula is a leafy green from the *Brassicaceae* family that goes by many different names, such as rocket, colewort, roquette, rucola and rucoli.

It has a slightly peppery taste and small leaves that can easily be incorporated into salads or used as a garnish. It can also be used cosmetically and medicinally.

Like other leafy greens, it's packed with nutrients such as vitamins A, B9 and K.

It's also one of the best sources of dietary nitrates, a compound that turns into 'nitric oxide in your body.

Though the benefits of nitrates are debated, some studies have found that they may help increase blood flow and reduce blood pressure by widening your blood vessels.

Endive

Endive belongs to the *Cichorium* family. It's less well known than other leafy greens, possibly because it's difficult to grow.

It's curly, crisp in texture and has a nutty and mildly bitter flavor. It can be eaten raw or cooked.

Just one-half cup (25 grams) of raw endive leaves packs 72% of the DV for vitamin K, 11% of the DV for vitamin A and 9% of the DV for folate.

It's also a source of kaempferol, an antioxidant that has been shown to reduce inflammation and inhibit the growth of cancer cells in test-tube studies.

Bok Choy

Bok choy is a type of Chinese cabbage.

It has thick, dark-green leaves that make a great addition to soups and stir-fries.

Bok choy is one of the few leafy green vegetables that contain the mineral selenium, which plays an important role in cognitive function, immunity and cancer prevention.

In addition, selenium is important for proper thyroid gland function. This gland is located in your neck and releases hormones that play a key role in metabolism.

An observational study associated low levels of selenium with thyroid conditions such as hypothyroidism, autoimmune thyroiditis and enlarged thyroid.

Turnip Greens

Turnip greens are the greens of the turnip plant, which is a root vegetable similar to potatoes.

These greens pack more nutrients than the turnip itself, including 'calcium, manganese, folate and the vitamins A, C and K.

They have a strong and spicy flavor and are often enjoyed cooked rather than raw.

Turnip greens are considered a cruciferous vegetable, which have been shown to decrease your risk of health conditions such as heart disease, cancer, inflammation and atherosclerosis.

Turnip greens also contain several antioxidants including gluconasturtiin, glucotropaeolin, quercetin, myricetin and beta-carotene — which all play a role in reducing stress in your body.

Turnip greens can be used as a replacement for kale or spinach in most recipes.

Microgreen

Frequently called "vegetable confetti," microgreens are young, tender greens that are used to enhance the color, texture, or flavor of salads, or to garnish a wide variety of main dishes. Harvested at the first true leaf stage and sold with the stem, cotyledons (seed leaves), and first true leaves attached, they are among a variety of novel salad greens available on the market that are typically distinguished categorically by their size and age. Sprouts, microgreens, and baby greens are simply those greens harvested and consumed in an immature state. Based on size or age of salad crop categories, sprouts are the youngest and smallest, microgreens are slightly larger and older (usually 2 in. tall), and baby greens are the oldest and largest (usually 3–4 in. tall).

Figure: Microgreens in this photo are predominantly in the cotyledon stage and are a few days away from harvest

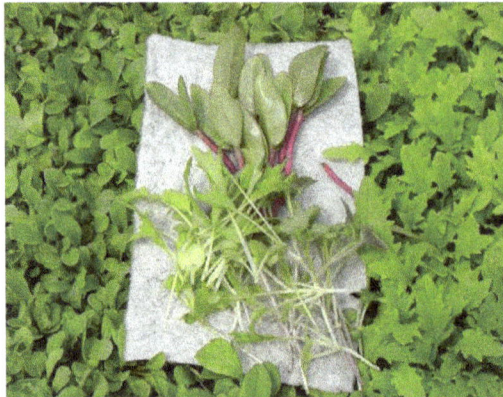

Figure: Microgreens are often termed "vegetable confetti"

Both baby greens and microgreens lack any legal definition. The terms "baby greens" and "microgreens" are marketing terms used to describe their respective categories. Sprouts are germinated seeds and are typically consumed as an entire plant (root, seed, and shoot), depending on the species. For example, sprouts from almond, pumpkin, and peanut reportedly have a preferred flavor when harvested prior to root development. Sprouts are legally defined, and have additional regulations concerning their production and marketing due to their relatively high risk of microbial contamination compared to other greens.

The crops used for microgreens usually do not include lettuces because they are too delicate and wilt easily. The kinds of crops that are selected for production and sale as microgreens have value in terms of color (like red or purple), unique textures, or distinct flavors. In fact, microgreens are often marketed as specialty mixes, such as "sweet," "mild," "colorful," or "spicy."

Certain crops of microgreens germinate easily and grow quickly. These include cabbage, beet, kale, kohlrabi, mizuna, mustard, radish, swiss chard, and amaranth. Soaking some seeds prior to sowing, such as beets, helps facilitate germination. As many as 80–100 crops and crop varieties have reportedly been used as microgreens. Others that have been used include carrot, cress, arugula, basil, onion, chive, broccoli, fennel, lemongrass, popcorn, buckwheat, spinach, sweet pea, and celery. Growers should evaluate various crop varieties to determine their value as microgreens. Many seed companies are very knowledgeable about the crops and varieties to grow, and a number of them offer organic seed.

Figure: A variety of crops can be grown and sold as microgreens

The commercial marketing of microgreens is mainly targeted toward restaurant chefs or upscale grocery stores. Prices for microgreens generally range from $30 to $50 per pound. The product is packaged in plastic clamshell containers that are typically 4–8 oz by weight but can be sold in 1 lb containers as well.

Production

Microgreens may be grown by individuals for home use. Growing small quantities at home is relatively easy; however, growing and marketing high-quality microgreens commercially is much more difficult. Having the right mix at the perfect stage for harvest is one of the most critical production strategies for success. The time from seeding to harvest varies greatly from crop to crop. When seeding a mixture of crops in a single planting flat, growers should select crops that have a similar growth rate so the entire flat can be harvested at once. Alternatively, growers can seed the various crops singularly and mix them after harvest.

Microgreens can be grown in a standard, sterile, loose, soilless germinating media. Many mixes have been used successfully with peat, vermiculite, perlite, coconut fiber, and others. Partially fill a tray with the media of choice to a depth of 1/2 in. to 1 or 2 in., depending on irrigation programs. Overhead mist irrigation is generally used only through the germination stage in these media systems. After germination, trays should be subirrigated to avoid excess moisture in the plant canopy.

An alternative production system uses one of several materials as a mat or lining to be placed in the bottom of a tray or longer trough. These materials are generally fiberlike and provide an excellent seeding bed. Materials may include burlap or a food-grade plastic specifically designed for microgreens such as those made by Sure to Grow. These mat systems are often used in a commercially available production system using wide NFT-type troughs. The burlap mat may be sufficient alone for certain crops or may require a light topping with a media after seeding. Seeding may be done as a broadcast or in rows. Seeding density is difficult to recommend. Most growers indicate they want to seed as thickly as possible to maximize production, but not too thickly because crowding encourages elongated stems and increases the risk of disease. Most crops require little or no fertilizer, as the seed provides adequate nutrition for the young crop. Some longer-growing microgreen crops, such as micro carrot, dill, and celery, may benefit from a light fertilization applied to the tray bottom. Some of the faster-growing greens, such as mustard cress and chard, may also benefit from a light fertilization because they germinate quickly and exhaust their self-contained nutrient supply quickly. Light fertilization is best achieved by floating each tray of microgreens for 30 seconds in a prepared nutrient solution of approximately 80 ppm nitrogen.

Microgreens are ready for harvest when they reach the first true leaf stage, usually at about 2 in. tall. Time from seeding to harvest can vary greatly by crop from 7 to 21 days. Production in small trays will likely require harvesting with scissors. This is a very time-consuming part of the production cycle and is often mentioned by growers as a major drawback. The seeding mat type of production system has gained popularity with many growers because it facilitates faster harvesting. The mats can be picked up by hand and held vertically while an electric knife or trimmer is used for harvesting, allowing cut microgreens to fall from the mat into a clean harvest container. Harvested microgreens are highly perishable and should be washed and cooled as quickly as possible. Some chefs are asking growers to deliver in the trays or mats and they will cut the microgreens as needed to improve quality. Wash the microgreens using good handling practices for food safety. Microgreens are usually packed in small, plastic clamshell packages and cooled to recommended temperatures for the crops in the mix. Growers should be aware that marketing agreements such as the National Leafy Green Marketing Agreement (NLGMA) have been proposed to reduce the risk of microbial contamination of mature and immature leafy greens.

Microgreens

Microgreens are young vegetable greens that are approximately 1–3 inches (2.5–7.5 cm) tall.

They have an aromatic flavor and concentrated nutrient content and come in a variety of colors and textures.

Microgreens are considered baby plants, falling somewhere between a sprout and baby green.

That said, they shouldn't be confused with sprouts, which do not have leaves. Sprouts also have a much shorter growing cycle of 2–7 days, whereas microgreens are usually harvested 7–21 days after germination, once the plant's first true leaves have emerged.

Microgreens are more similar to baby greens in that only their stems and leaves are considered edible. However, unlike baby greens, they are much smaller in size and can be sold before being harvested.

This means that the plants can be bought whole and cut at home, keeping them alive until they are consumed.

Microgreens are very convenient to grow, as they can be grown in a variety of locations, including outdoors, in greenhouses and even on your windowsill.

Different Types of Microgreens

Microgreens can be grown from many different types of seeds.

The most popular varieties are produced using seeds from the following plant families:

- Brassicaceae family: Cauliflower, broccoli, cabbage, watercress, radish and arugula
- Asteraceae family: Lettuce, endive, chicory and radicchio
- Apiaceae family: Dill, carrot, fennel and celery
- Amaryllidaceae family: Garlic, onion, leek
- Amaranthaceae family: Amaranth, quinoa swiss chard, beet and spinach
- Cucurbitaceae family: Melon, cucumber and squash

Cereals such as rice, oats, wheat, corn and barley, as well as legumes like chickpeas, beans and lentils, are also sometimes grown into microgreens.

Microgreens vary in taste, which can range from neutral to spicy, slightly sour or even bitter, depending on the variety. Generally speaking, their flavor is considered strong and concentrated.

Microgreens are Nutritious

Microgreens are packed with nutrients.

While their nutrient contents vary slightly, most varieties tend to be rich in potassium, iron, zinc, magnesium and copper.

Microgreens are also a great source of beneficial plant compounds like antioxidants.

What's more, their nutrient content is concentrated, which means that they often contain higher vitamin, mineral and antioxidant levels than the same quantity of mature greens.

In fact, research comparing microgreens to more mature greens reports that nutrient levels in microgreens can be up to nine times higher than those found in mature greens.

Research also shows that they contain a wider variety of polyphenols and other antioxidants than their mature counterparts.

One study measured vitamin and antioxidant concentrations in 25 commercially available microgreens. These levels were then compared to levels recorded in the USDA National Nutrient Database for mature leaves.

Although vitamin and antioxidant levels varied, levels measured in microgreens were up to 40 times higher than those recorded for more mature leaves.

That said, not all studies report similar results.

For instance, one study compared nutrient levels in sprouts, microgreens and fully grown amaranth crops. It noted that the fully grown crops often contained as much, if not more, nutrients than the microgreens.

Therefore, although microgreens generally appear to contain higher nutrient levels than more mature plants, this may vary based on the species at hand.

Health Benefits of Microgreens

Eating vegetables is linked to a lower risk of many diseases.

This is likely thanks to the high amounts of vitamins, minerals and beneficial plant compounds they contain.

Microgreens contain similar and often greater amounts of these nutrients than mature greens. As such, they may similarly reduce the risk of the following diseases:

- Heart disease: Microgreens are a rich source of polyphenols, a class of antioxidants linked to a lower risk of heart disease. Animal studies show that microgreens may lower triglyceride and "bad" LDL cholesterol levels.

- Alzheimer's disease: Antioxidant-rich foods, including those containing high amounts of polyphenols, may be linked to a lower risk of Alzheimer's disease.

- Diabetes: Antioxidants may help reduce the type of stress that can prevent sugar from properly entering cells. In lab studies, fenugreek microgreens appeared to enhance cellular sugar uptake by 25–44%.

- Certain cancers: Antioxidant-rich fruits and vegetables, especially those rich in polyphenols, may lower the risk of various types of cancer. Polyphenol-rich microgreens may be expected to have similar effects.

Storage and Commercial Transport

Microgreens have a short shelf life and better methods of storing and transporting microgreens are currently being studied, which at this time are mainly focusing on buckwheat. Commercial microgreens are most often stored in plastic clamshell containers, which do not provide the right balance of oxygen and carbon dioxide for any live greens to breathe. Among package materials called films, differences in permeability are referred to as the oxygen transmission rate.

The ARS researchers found that buckwheat microgreens packaged in films with an oxygen transmission rate of 225 cubic centimeters per square inch per day had a fresher appearance and better cell membrane integrity than those packaged in other films tested. Following these steps, the team maintained acceptable buckwheat microgreen quality for more than 14 days—a significant extension, according to authors.

Light-emitting diodes. otherwise known as LEDs, now provide the ability to measure impacts of narrow-band wavelengths of light on seedling physiology. The carotenoid zeaxanthin has been hypothesized to be a blue light receptor in plant physiology. A study was carried out to measure the impact of short-duration blue light on phytochemical compounds, which impart the nutritional quality of sprouting broccoli microgreens. Broccoli microgreens were grown in a controlled environment under LEDs using growing pads. Short-duration blue light acted to increase important phytochemical compounds influencing the nutritional value of broccoli microgreens.

Spinach

Obscurely referred to for years as "the Spanish vegetable" in England, the name of this leafy green was later shortened to what we call it today.[1] Spinach cultivation is thought to have originated from ancient Persia, later spreading to Nepal, and by the seventh century, to China, where it's still called "Persian Greens." The Moors introduced it to Spain around the 11th century.

According to the United States Department of Agriculture (USDA), Americans consume nearly 2 1/2 pounds of spinach per year, per capita. This easily quadruples the amount eaten 40 years ago, possibly because of a drastic overhaul in image and presentation.

Now greener, tastier and crisper by freezing, fresh spinach is often used for salads and in place of lettuce on sandwiches.

No other vegetable has ever gained the fame that spinach did in the 193 0s through the cartoon character Popeye. Parents often encouraged their children to eat their spinach so they would grow up to be big and strong like the cartoon sailor, and there's actually some truth to this claim.

Health Benefits of Spinach

- Spinach is storehouse for many phytonutrients that have health promotional and disease prevention properties.

- It is very low in calories and fats (100 g of raw leaves provide just 23 calories). Also, its leaves hold a good amount of soluble dietary fiber; no wonder why this leafy greens often recommended by dieticians in the cholesterol controlling and weight reduction programs!

- Fresh 100 g of spinach contains about 25% of daily intake of iron, one of the highest for any green leafy vegetables. Iron is an essential trace element required by the human body for red blood cell production and as a co-factor for an oxidation-reduction enzyme, cytochrome oxidase during the cellular metabolism.

- Fresh leaves are a rich source of several vital antioxidant vitamins like vitamin-A, vitamin-C, and flavonoid polyphenolic antioxidants such as *lutein, zeaxanthin, and β-carotene*. Together, these compounds help act as protective scavengers against oxygen-derived free radicals and reactive oxygen species (ROS) that play a healing role in aging and various disease processes.

- *Zeaxanthin*, an important dietary carotenoid, is selectively absorbed into the retinal macula lutea in the eyes where it thought to provide antioxidant and protective UV light-filtering functions. It thus helps protect from "age-related macular related macular disease" (ARMD), especially in the older adults.

- Further, vitamin-A is required for maintaining healthy mucosa and skin and is essential for night vision. Consumption of natural vegetables and fruits rich in vitamin-A and flavonoids are also known to help the body protect from lung and oral cavity cancers.

- Spinach leaves are an excellent source of vitamin-K. 100 g of fresh greens provides 402% of daily vitamin-K requirements. Vitamin-K plays a vital role in strengthening the bone mass by promoting osteoblastic activity in the bones. Additionally, it also has an established role in patients with *Alzheimer's disease* by limiting neuronal damage in the brain.

- This green leafy vegetable also contains good amounts of many B-complex vitamins such as vitamin-B6 (pyridoxine), thiamin (vitamin B-1), riboflavin, folates, and niacin. Folates help prevent neural tube defects in the newborns.

- 100 g of farm fresh spinach has 47% of daily recommended levels of vitamin-C. Vitamin-C is a powerful antioxidant, which helps the body develop resistance against infectious agents and scavenge harmful oxygen-free radicals.

- Its leaves also contain a good amount of minerals like *potassium*, manganese, magnesium, copper and zinc. Potassium is an important component of cell and body fluids that helps controlling heart rate and blood pressure. The human body uses manganese and copper as a co-factor for the antioxidant enzyme, *superoxide dismutase*. Copper is also required for the production of red blood cells. Zinc is a co-factor for many enzymes that regulate growth and development, digestion and nucleic acid synthesis.

- It is also a small source of omega-3 fatty acids.

Regular consumption of spinach in the diet helps prevent osteoporosis.

Planting

- Prepare the soil with aged manure about a week before planting, or, you may wish to prepare your spot in the fall so that you can sow the seeds outdoors in early spring as soon as the ground thaws.

- If you live in a place with mild winters, you can also plant in the fall.

- Although seedlings can be propagated indoors, it is not recommended, as seedlings are difficult to transplant.

- Spring plantings can be made as soon as the soil can be properly worked. It's important to seed as soon as you can to give spinach the required 6 weeks of cool weather from seeding to harvest.

- Select a site with full sun to light shade and well-drained soil.

- Sow seeds ½ inch to 1 inch deep, covering lightly with soil. Sow about 12 seeds per foot of row, or sprinkle over a wide row or bed.

- Soil should not be warmer than 70° F in order for germination.

- Successive plantings should be made every couple weeks during early spring. Common spinach cannot grow in midsummer. (For a summer harvest, try New Zealand Spinach or Malabar Spinach, two similar leafy greens.)

- Plant in mid-August for a fall crop, ensuring that soil temps are cool enough.

- Gardeners in northern climates can harvest early-spring spinach if it's planted just before the cold weather arrives in fall. Protect the young plants with a cold frame or thick mulch through the winter, then remove the protection when soil temperature in your area reaches 40°.

- Water the new plants well in the spring.

Care

- Fertilize only if necessary due to slow growth, or use as a supplement if your soil pH is inadequate. Use when plant reaches ⅓ growth.

- When seedlings sprout to about two inches, thin them to 3-4 inches apart.

- Beyond thinning, no cultivation is necessary. Roots are shallow and easily damaged.

- Keep soil moist with mulching.

- Water regularly.

- Spinach can tolerate the cold; it can survive a frost and temps down to 15° F.

Harvesting and Storage

Harvest regularly to stimulate regrowth and a higher yield that won't bolt (go to seed) easily. Harvest by taking off the outer leaves with a sharp knife 30mm above the soil.

Under normal conditions the plant can be grown for five months before new plants are sown or transplanted.

If the leaves are not for immediate use, bunch them and place them in water to keep them fresh for longer.

Because of the high transpiration rate of the broad leaves, the keeping quality is poor. Stored in a fridge or cold store, spinach will keep for six to eight days.

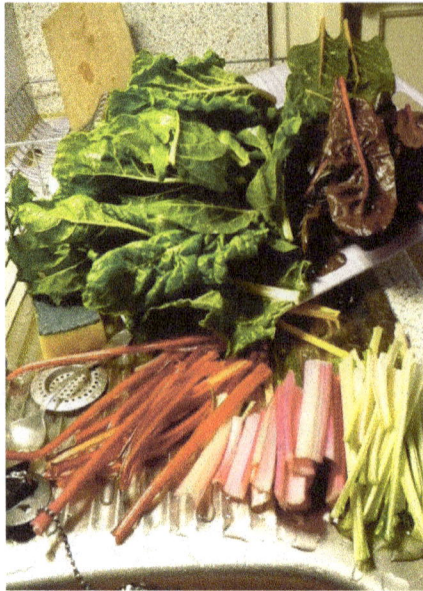

The leaves and the stems are edible and may be marketed and cooked separately. Swiss chard is a good source of vitamin A, magnesium and iron. Vitamin A stops night blindness and helps with chronic fatigue and heart disease.

Pests and Disease

Spinach may suffer from a number of diseases. The best way to avoid this is to keep the plants healthy by providing the right nutrition, enough water and a good start.

Monitor the crop and keep scouting for pests, removing them as you see them.

For aphids, and red spider mite, a weak solution of soapy water (you can use Sunlight liquid) will do the trick.

Cutworms hide under the soil and come out at night to feed. They cut the seedlings down at the soil/air interface. Use cutworm bait and keep checking.

To keep diseases and pests at bay, rotate and intercrop with marigold, mustard and rapeseed and leave the land fallow.

Sissoo Spinach

Sissoo spinach, also known as Brazilian spinach, is a tropical edible groundcover of the genus Alternanthera and used as a leaf vegetable. Sissoo spinach is a vigorous and spreading groundcover about 30 cm high with crinkly leaves, rooting at the nodes. It's great to use as a garden border or plant underneath fruit trees to hold in moisture. If you can plant many starts at a time, it will provide abundant greens all year round.

One need not worry as it's not considered an invasive species.

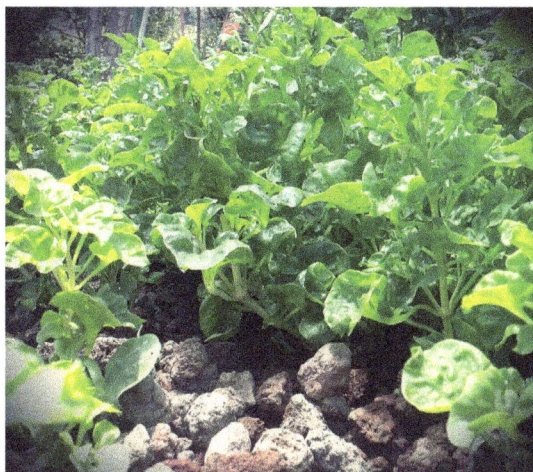

A spinach that thrives in hot climates

The leaves are crunchy, and not slimy. It's eaten raw or added to cooked dishes as a spinach substitute. Brazilians commonly eat it raw in salads with oil and or vinegar, tomato, and onion. However, the leaves need steaming or boiling when eaten in large quantities because of the presence of oxalates.

Methods to Grow Sissoo Spinach

Sissoo doesn't go to seed, however, it is easily propagated from cuttings. It will create a mound on the ground that requires regular trimming/eating. It prefers 50% shade and tolerates a wide range of pH soil conditions. Since its main product is its leaves, it needs a high amount of nitrogen, organic matter and water. We mulch around it and line the beds with compost. It can grow in full sun but will do better with a little shade. Plants are prone to leaf-eating caterpillar pests and slugs. We deal with these using a chili garlic detergent water mix. It can also be planted as a living mulch under fruit trees to help hold moisture in the soil.

If you have an aquaponics system, Sissoo is a must have. The high nitrogen environment make growing and propagating a breeze. When we harvest, we just make a few cutting and put them back in the rock grow beds. Within a week they develop hearty root systems.

Celery

Celery is a long-season vegetable grown in the spring or fall. It has the reputation of being fussy; however, it's really quite easy if you understand its specific needs.

Celery has three critical needs:

1. A long growing season (130 to 140 days of mostly cool weather). Celery will not tolerate high heat.

2. A constant, unfailing water supply. The soil must stay watered at all times. If celery has a spell without water, it will be problematic (stringy, tough, and/or hollow stalks).

3. Rich, fertile soil with plenty of organic matter mixed in. The crop is a big feeder and also needs to be fertilized during its growth period, too. Because celery roots are shallow (just a few inches deep), make sure nutrients are in the top of the soil.

Celery is often grown as a winter crop in the South, a summer crop in the far North, and a fall crop in most other areas.

Transplants are hard to find, so be prepared to start plants from seed.

Planting

- Because the season is so long, celery seeds should always be started indoors for the best success rate. For a spring crop, sow seeds indoors 10 to 12 weeks before the average last frost date for your area. For a fall crop, sow in summer, timing so that you can set out transplants 10 to 12 weeks before the first autumn frost.

- Soak seeds in warm water overnight prior to planting to reduce germination time.

- To get good germination, don't cover the seeds with soil. Simply press the seeds into potting soil that's formulated for seed-starting and cover the trays or pots with plastic covers to hold the moisture. Germination should take place in about a week

- When the plants are two inches tall, transplant them to individual peat pots or to another, deeper, flat with new potting soil. If you use flats, put the plants at least two inches apart.

- Plant celery outdoors when the soil temperature reaches 50 degrees F. or more, and when the nights don't dip down below 40 degrees F.

- Work organic compost into the soil prior to planting. Or mix in fertilizer (about one pound of 5-10-10 per 30 square feet).

- Harden off seedlings before transplanting by reducing water slightly, and keeping them outdoors for a couple hours a day.

- Transplant seedlings 8 to 10 inches apart Direct sow seeds ¼ inch deep. These will need to be thinned to 12 inches apart when they reach about six inches high.

- Mulch the plants after they are 6 inches tall to keep the soil moist and the roots cool.

- Water directly after planting.

Care

- Celery requires lots of water. Make sure to provide plenty of water during the entire growing season, especially during hot, dry weather. If celery does not get enough water, the stalks will be dry and small.

- Add plenty of compost and mulch around the plants to retain moisture. Sidedress with a 5-10-10 fertilizer in the second and third month of growth (one tablespoon per plant and sprinkle it in a shallow furrow three to four inches from the plant and cover it with soil).

- Keep celery weeded but be careful when weeding as celery has shallow roots and could easily get distrubed.

- Tie growing celery stalks together to keep them from sprawling.

Pests/Diseases

- Flea beetles

- Slugs and snails

- Earwigs

To control pests, cover the plants with garden fabric (row covers) during the first four to six weeks of the growing season.

Harvest/Storage

- The parts of celery that are harvested are mainly the stalks, which will be above ground.

- Pick the stalks whenever you want. Young celery is as good as the mature product.

- Harvest stalks from the outside in. You may begin harvesting when stalks are about 8 inches tall.

- Celery can be kept in the garden for up to a month if soil is built up around it to maintain an ideal temperature. Celery will tolerate a light frost, but not consecutive frosts.

- Tip: The darker the stalks become, the more nutrients they will contain. Texture changes with color; dark green stalks will be tougher.

- Keep celery in a plastic bag in the refrigerator. Celery stores really well; you can keep it for many weeks with no trouble.

Chinese Cabbage

Chinese cabbage is a leafy vegetable native to china province. It is also being grown in other Asian countries. Chinese cabbage is a leading market vegetable in China, Japan, Korea and Southeast Asian region. Chinese cabbage is closely related to vegetables like broccoli, or cauliflower. Chinese cabbage can be used in many dishes such as boiled in soups, salads or stir fries. In India, Fresh leaves can be used for curry/chutney/leaf fry. Though this cabbage is called with different names in Asian countries, outside world knows this vegetable as "Chinese Cabbage". The Chinese cabbage crop is a cool season annual vegetable crop which thrives best when the days are short and mild. Chinese cabbage tastes mild and aromatic and it is good source of calcium and vitamin 'C'. If you are looking for low calorie leafy vegetable, this cabbage is perfect one for you. When it comes to vegetable and plant description, Chinese cabbage inner leaves are in light yellow colour and this plant is like long (oblong) shaped head with tightly contained crinkly, thick, light green leaves with white notable veins that can grow up to 15-30 cm tall. Fresh leaves and tender shoots are essential parts of this plant. Commercial cultivation of Chinese cabbage is very much profitable as this leafy vegetable has very good demand in the local markets especially in Asian countries.

Climate Requirement for Chinese Cabbage Farming

Chinese cabbage thrives best during the cooler periods of the growing season. It prefers an average temperature of 15°C to 22°C during crop early growth. Temperatures below 0°C are tolerated for short time periods but too low temperature can induce premature bolting of the crop. Chinese cabbage is a cool seasonal crop and needs good amount of water supply throughout plant growth period. However, The most critical stage of watering is during cabbage head formation. This crop grows best in winter and should have proper irrigation for good yields.

Soil Requirement

Chinese cabbage can be grown on wide variety of soils ranging from sandy loam to textured loam. However, well-drained sandy loam soils with good organic matter are proved to be good for excellent yield and quality produce. The idea pH range for Chinese cabbage farming is 5.5 – 7.0. Lower pH soils always lead to calcium or magnesium deficiency hence they must be compensated with required nutrients. Avoid soils which are extremely sandy and claylike. As this crop requires plenty of water, a good moisture holding soil is very well recommended for growing the Chinese cabbage.

Propagation

Chinese cabbage propagation is done from seed. Seedling can be raised on nursery beds or greenhouse and can be transplanted to main field. Even they can be directly sown on field beds.

Land Preparation

Land should be prepared well enough to have fine tilth of soil. Soil can be tilled by using fork or hoe. Land should be prepared at least 2 months in advance of planting. Chisel Plough or Disc plough can be used to prepare the land. Any weeds or any plant residues should be removed in order to make field weed free. In case of sandy soils, rotary hoed cultivation is preferred to make a good seedbed.

Seed Rate, Planting and Spacing

Chinese Cabbage Seeds

With the single grain seed technique, about 500–600 grams of seed/hectare is required, for normal and traditional seed technique about 1 kg of seed / hectare is recommended. If traditional or normal seed technique is practiced, the seedlings must be thinned out after 2 to 4 weeks after sowing. In case of direct seeding in the field, make a planting furrow about 2-3 cm using a hand hoe. Planting should be done in seedbeds, in row spaced 15 to 20 cm distance. Seeds should be sown 1-2 cm apart within the row. Seeds should be sprinkled into the furrow and covered with 1 to 1.5 cm of soil.

In case of transplantation of seedlings, Chinese cabbage seedlings are grown in trays which hold 100 to 125 plants in nurseries or greenhouse and the individual cell-pack may have a diameter of only 15 mm and a depth of 10 mm. Seedlings can transplant when the seedlings are about 15 to 16 cm tall or 2 to 3 weeks after sowing. Generally it requires 75,000 to 80,000 seedlings / hectare and the transplanting method is used for the spring crop and the seeding technique is used for the fall crop.

Irrigation

Well, Chinese cabbage crop requires good irrigation throughout its lifecycle or growth period. The frequency of irrigation actually depends on soil type, climate , and plant or crop age. However, in case of Napa/Chinese cabbage it is better to irrigate 3 times a week for sandy soils, if the soil is sandy loam type, it may require 2 times a week. make sure to maintain constant moisture at plant roots. Irrigation is very critical and any water stress should be avoided at head forming stage. In case of heavy rains, make sure to drain out the water from the field as early as possible. Drip irrigation, flood irrigation or micro jet sprinkler irrigation can be adopter for better water utilization.

Manures and Fertilizers

This crop responds very well to manures and fertilizers. Application of nutrients is very important for quality produce and high yields.

15 to 20 tonnes of Well-decomposed farm yard manure (FMY) per 1 hectare field should be applied during land preparation. The other fertilizers details are as follows.

- Nitrogen (N): 160–200 kg/ha
- Phosphorus (P_2O_5): 80–120 kg/ha
- Potash (K_2O): 180–250 kg/ha
- Calcium (C) : 100–150 kg/ha
- Magnesium (Mg): 20–40 kg/ha

Best results can be achieved when the nitrogen (N) is broadcasted before planting, and also side-dressed in 1 or more applications 10 to 12 days apart following thinning or within one month of transplanting in the field.

Intercultural Operations

Weed control is very important in any successful horticulture crops. Weeds can be controlled by hand (hand hoeing or hand pulling), machine or herbicides / weedicides. Mulching is another practice of checking weeds growth which is also useful in keeping moisture intact. Earthing up should be done after 2 months of planting. shallow cultivation may be given in the rows to check any growing weeds.

Pests and Diseases

The following are the pests and disease found in Chinese Cabbage Farming.

- Pests and Control Measures: The main pests in Chinese cabbage farming are Bagrada bugs, Cutworms, Aphids, and Spider mites. For symptoms and control measures, contact any technical officer in horticulture department. These Pests can also be controlled by removing weeds, affected plants and keeping the field clean.

- Diseases and Control Measures: The most common and serious diseases affecting Chinese cabbage crop are downy mildew, powdery mildews, club root and black rot. These diseases can be controlled.

- By using clean seeds or removing all infected plants. For symptoms and chemical control of these diseases, contact your local horticulture department.

Harvesting

Chinese cabbage becomes ready for harvesting within 70 to 100 days after planting depending variety and other factors. The Chinese cabbage leaves can be harvested by hand when the leaves are fully developed. The leaves should be cut (dislodged) at the plant base. In general, harvesting can be done when the plant reaches the 8-leaf stage and usually this happens after 70 days to 100 days after planting.

Harvested Chinese Cabbage

Post-harvesting

The following activities should be performed after harvesting the Chinese cabbage crop.

1. Cleaning: The cabbage leaves should be washed to remove the soil or any dirt immediately after harvesting from the field.

2. Storage: As the shelf life of Chinese cabbage is short , one should make sure to store the leaves in cool place.

3. Transport: Chinese cabbage requires cooling transport to retain the turgidity of the leaves.

4. Marketing: Local markets or any vegetable vendors are best source of marketing. Even you can sell fresh leaves at farm gate. Chinese cabbage should be sold within 2 days after harvesting.

Yield

Any agriculture or horticulture crop yield depends on many factors like cultivar (variety), soil type, climate and crop cultivation practices. On an average one can obtain a yield of 4 to 5 kg/ sq. meter plantation.

Napa Cabbage

Napa cabbage is one of the favorite leafy-cabbage vegetables in mainland China. Napa's sweet, crunchy, and celery-flavored leaves are one of the most sought-after ingredients in the far East-Asian cuisine. Undoubtedly, Chinese cabbages are increasingly being used in the western, Mediterranean as well as American cuisines for their wholesome nutrition profile.

Botanically, this chinese cabbage variety belongs to the *Brassica* family; a large class of leafy/flower-head vegetables which also includes brussels sprouts, kale, cabbage, and broccoli, etc. Scientific name: B. campestris (Pekinensis group). Some of the common names of napa are pe-tsai, 白菜, celery cabbage, Chinese white cabbage, Peking cabbage, won bok, nappa (Japanese), hakusai (Japanese), pao, hsin pei tsai, kimchi cabbage, etc.

Fresh napa-cabbage

Napa cabbage is an annual, cool season vegetable. It grows best when the days are short and mild. As in cabbages, napa grows to oblate shaped heads consisting of tightly arranged crinkly, thick, light-green leaves with prominent, pale white veins. At its core, the leaves feature smooth, light-yellow hue.

Napa cabbage

There exist two major types of napa cabbage; Chilili and Che foo. *Chilili* types produce cylindrical heads, measuring about 18 inches long and 6 inches wide, featuring erect, upright growing habit. *Che-foo* type forms compact, round head of green-blade with white-petioled leaves.

Health Benefits of Napa Cabbage

- Napa cabbage is incredibly low in calories. 100 g fresh leaves carry jus 16 calories. Along with celery, bok-choy, etc., it easily fits into the neo-class of zero calorie or negative calorie group of vegetables as often advocated by some dieticians.

- Napa packed with many antioxidant plant compounds such as *carotenes, thiocyanates, indole-3-carbinol, lutein, zeaxanthin, sulforaphane* and *isothiocyanates*. Also, it is an abundant source of soluble and insoluble dietary fiber. Scientific studies suggest these compounds are known to offer protection against breast, colon and prostate cancers and help reduce LDL or "bad cholesterol" levels in the blood.

- Fresh napa is an excellent source of folates. 100 g provides 79 µg or 20% of daily required levels of this B-complex vitamin. Folic acid is one of the essential components of DNA. Sufficient amounts of folates in the diet in anticipant mothers may help prevent neurological diseases in the newborn babies.

- Further, Napa cabbage has great levels of vitamin-C. 100 g of fresh napa provides about 45% of daily requirements of this vitamin. Regular consumption of foods rich in vitamin-C helps the body develop resistance against infectious agents and scavenge harmful, pro-inflammatory free radicals.

- Likewise in other cabbages, napa too has moderate levels of vitamin-K, provides about 38% of RDA levels. Vitamin-K has a potential role in the bone metabolism by promoting osteoblastic activity in bone cells. Therefore, sufficient levels of vitamin K in the diet makes the bone stronger, healthier and help delay osteoporosis. Further, vitamin-K also has established role to play in the treatment of Alzheimer's disease patients by limiting neuronal damage in their brain.

- Napa cabbage has small levels of vitamin-A. However, it also contains flavonoid polyphenolic compounds such as carotenes, lutein, and xanthin which convert to vitamin-A in the human body.

- As in other green vegetables, it is a good source of many essential vitamins such as riboflavin, pantothenic acid, pyridoxine (185 of RDA) and thiamin. These vitamins are essential in the sense that our body requires them from external sources to replenish.

- Also, it is a very natural source of electrolytes and minerals like calcium, potassium, phosphorous, manganese, iron and magnesium. Potassium is the chief component of cell and body fluids and helps in regulating heart rate and blood pressure. The human body employs manganese as a co-factor for the antioxidant enzyme, *superoxide dismutase*. Iron is essential for the red blood cell formation.

Cultivation

Napa cabbage can be cultivated in many different areas of the world, the main area of diversification represents Asia.

Soil Requirements

Napa cabbage requires deeply loosened medium heavy soil. There must not be any compaction due to plowing. The crop achieves particularly high yields on sandy loam. Extremely sandy or claylike soils are not suitable. The crop prefers a pH range from 6.0 to 6.2, a high organic matter content and good moisture holding capacity of the soil. Lower pH or droughty soil can lead to calcium or magnesium deficiency and internal quality defects.

Climate Requirements

Napa cabbage needs much water during the whole growth period. Often an irrigation system is needed, especially for August and September. The required amount of water depends on the stage of crop growth, weather conditions, and soil type. The most critical stage after establishment is when the head is forming. Inadequate water at this time will result in reduced uptake of calcium. This condition causes dead leaf tips within the head what makes it unmarketable. During head formation, 1 to 1 ½ inches of water per week is needed to maintain sustained growth rates.

Temperature requirements are low. Temperatures below zero degrees are tolerated for short time periods; persistent frosts below -5°C are not endured. Too low temperature can induce premature bolting. The plants perform best under temperatures between 13°C and 21°C, but depending on the cultivar.

Seedbed Requirements and Sowing

Napa cabbage has very small seeds with a thousand kernel weight of about 2.5–2.8 g. For professional cultivation it is recommended to use disinfected seeds to prevent onset diseases. With the single grain seed technique, about 400–500 g seeds per hectare is required, for normal seed technique about 1 kg per hectare. If normal seed technique is used, the seedlings must be thinned out after 2–4 weeks. The seeds should be deposited 1–2 cm deep, with a row width of 40–45 cm and 25–30 cm distance between the seeds.

The seedlings can be grown in the greenhouse and then transplanted into the field after 2–3 weeks. Earlier harvest can be achieved with this method. Seventy thousand to 80,000 seedlings per hectare

are required. The transplanting method is normally used for the spring crop and the seeding technique for the fall crop.

Fertilization, Field Management

The nutrient removal of napa cabbage is high:

- 150–200 kg N per hectare
- 80–120 kg P_2O_5 per hectare
- 180–250 kg K_2O per hectare
- 110–150 kg Ca per hectare
- 20–40 kg Mg per hectare

Fertilizer recommendations are in the range of the nutrient removal. Organic fertilizer must be applied before sowing due to the short cultivation time of napa cabbage and the slow availability of organic fertilizers. Synthetic N fertilizer should be applied in 3 equal doses. The last application must happen before 2/3 of the cultivation time is over to avoid quality losses during storage.

Weeds should be controlled mechanically or chemically.

Harvest, Storage and Yield

Harvested napa cabbage being loaded on a truck in Tonghai County

Napa cabbage can be harvested 8–12 weeks after sowing. The harvest work is mostly done by hand. The plant is cut 2.5 cm above the ground. It is usual to harvest several times per field to achieve constant quality of the napa cabbage. Storage of napa cabbage is possible for 3–4 months in cool stores with 0-1°C and 85-90 percent relative humidity. Napa cabbage achieves a yield of 4–5 kg per m².

Breeding

Brassica rapa species are diploid and have 10 chromosomes. A challenge for breeding of napa cabbage is the variable self-incompatibility. The self-incompatibility activity was reported to change by temperature and humidity. In vitro pollination with 98% relative humidity proved to be the most reliable as compared to greenhouse pollination.

A lot of work has already been done on breeding of napa cabbage. In the 21st century, 880 varieties of Napa cabbage were registered by the Korea Seed and Variety Service.

Breeding of napa cabbage was started by the Korean government research station of horticultural demonstration in 1906 to overcome starvation. As napa cabbage and radish are the main vegetables for "Kimchi" research focused on increasing yield. The most important person for this process was Dr. Woo Jang-choon who bred hybrid cultivars with self-incompatibility and contributed to commercial breeding by developing valuable materials and educating students. The main purpose of the hybrid cultivar was high yield and year-round production of napa cabbage after 1960.

To enable year round production of napa cabbage, it has to be modified to tolerate high and low temperatures. Normally sowing in the late summer and harvesting in late autumn can produce high quality vegetables. As an example, a summer cultivar called "Nae-Seo-beak-ro" was developed 1973 by a commercial seed company. It tolerates high temperatures, could endure high humidity in the monsoon, and showed resistance to viral disease, soft rot and downy mildew. The low temperature in early spring reduces the quality of the vegetable and it cannot be used for "Kimchi". In the 1970s the developing of winter cultivars started. The majority of new cultivars could not endure the cold winter conditions and disappeared. The cultivar "Dong-Pung" (meaning "east wind") was developed 1992 and showed a high resistance to cold temperature. It is mostly used in Korea, where fresh napa cabbage is nowadays cultivated all year round.

In the 70s, one seed company developed the rose-shape heading variety while other seed companies focused on semi-folded heading type. As a result of continuous breeding in the commercial seed companies and the government research stations, farmers could now select what they wanted from among various high quality hybrids of Chinese cabbage.

In 1988, the first cultivar with yellow inner leaf was introduced. This trait has prevailed until today.

A very important breeding aim is to get varieties with resistance to pests and diseases. There exist varieties with resistance to turnip mosaic virus but as mentioned above, there exist numerous other diseases. There have been attempts to breed varieties with clubroot resistance or powdery mildew resistance but the varieties failed due to bad leaf texture traits or broken resistances.

Pests and Diseases

Fungal Diseases

Alternaria diseases are caused by the organisms *Alternaria brassicae*, *Alternaria brassicicola* and *Alternaria japonica*. Their English names are black spot, pod spot, gray leaf spot, dark leaf spot or Alternaria blight. The symptoms can be seen on all aboveground plant parts as dark spots. The infected plants are shrivelled and smaller than normal. Alternaria diseases infect almost all brassica plants, the most important hosts are oilseed brassicas. The fungus is a facultative parasite, what means that it can survive on living hosts as well as on dead plant tissue. Infected plant debris is in most circumstances the primary source of inoculum. The spores can be dispersed by wind to host plants in the field or to neighbouring brassica crops. This is why cross infections often occur in areas where different brassica crops are cultivated in close proximity. The disease spreads especially fast when the weather is wet and the plants have reached maturity. *Alternaria brassicae* is well adapted to temperate regions while *Alternaria brassicicola* occurs primarily in warmer parts of the world.

Temperature requirement for *Alternaria japonica* is intermediate. There exist some wild accessions of *Brassica rapa* subsp. *pekinensis* with resistance to *Alternaria brassicae* but not on commercial cultivars. These resistances should be included to breeding programmes. *Alternaria* epicemics are best avoided by management practices like at least 3 years non-host crops between brassica crops, incorporation of plant debris into the soil to accelerate decomposition and usage of disease-free seeds.

Anhracnose is a brassica disease caused by *Colletotrichum higginsianum* that is especially damaging on napa cabbage, pak choi, turnip, rutabaga and tender green mustard. The symptoms are dry pale gray to straw spots or lesions on the leaves. The recommended management practices are the same as for *Alternaria* diseases.

Black root is a disease that infects mainly radish, but it also occurs on many other brassica vegetables inclusively napa cabbage. It caused by the fungus *Aphanomyces raphani*. The pathogen can persist for long times in the soil, therefore crop rotations are an essential management tool.

White leaf spot is found primarily in temperate climate regions and is important on vegetable brassicas and oilseed rape. The causal organism is *Mycosphaerella capsellae*. The symptoms are white spots on leaves, stems and pods and can thus easily be confused with those of downy mildew. The disease spreads especially fast with rain or moisture and temperature is between 10 and 15°C.

Yellows, also called Fusarium wilt, is another brassica disease that infects oilseed rape, cabbage, mustards, Napa cabbage and other vegetable brassicas. It is only a problem in regions with warm growing seasons where soil temperatures are in the range of 18 to 32°C. The causal organism is *Fusarium oxysporum* f. sp. *conlutinans*. Napa cabbage is relatively tolerant to the disease; mostly the only external symptoms are yellowing of lower, older leaves. The disease is soil borne and can survive for many years in the absence of a host. Most cruciferous weeds can serve as alternate hosts.

Damping-off is a disease in temperate areas caused by soil inhabiting oomycetes like *Phytophthora cactorum* and *Pythium* spp. The disease concerns seedlings, which often collapse and die.

Links to other diseases that infect napa cabbage:

- Black leg or phoma stem cancer: *Leptosphaeria maculans*
- Clubroot: *Plasmodiophora brassicae*
- Downy mildew: *Hyaloperonospora brassicae*
- Powdery mildew: *Erysiphe cruciferarum*
- Rhizoctonia solani
- Sclerotinia sclerotiorum.

Bacterial Diseases

Bacterial soft rot is considered one of the most important diseases of vegetable brassicas. The disease is particularly damaging in warm humid climate. The causal organisms are *Erwinia carotovora* var. *carotovora* and *Pseudomonas marginalis* pv. *marginalis*. The rot symptoms can occur in the field, on produce transit or in storage. Bacteria survive mainly on plant residues in the soil. They are spread by insects and by cultural practices, such as irrigation water and farm machinery.

The disease is tolerant to low temperatures; it can spread in storages close to 0°C, by direct contact and by drippint onto the plants below. Bacterial soft rot is more severe on crops which have been fertilized too heavily with nitrogen, had late nitrogen applications, or are allowed to become over-mature before harvesting.

Black rot, the most important disease of vegetable brassicas, is caused by *Xanthomonas campestris* pv. *campestris*.

Virus Diseases

- Cucumber mosaic virus
- Radish mosaic virus
- Ribgrass mosaic virus
- Turnip crincle virus
- Caradamine chlorotic fleck virus
- Turnip mosaic virus
- Turnip yellow mosaic virus.

Insect Pests

- Large white butterfly *(Pieris brassicae)*
- Cabbage root fly *(Delia radicum)*
- Cabbage seed weevil *(Ceutorhynchus assimilis)*
- Cabbage looper
- Diamondback moth
- Small white butterfly *(Pieris rapae)*
- Aphids
- Cucumber beetles
- Stink bugs
- Vegetable weevils
- Mole crickets
- Cutworms.

Other Pests and Diseases

Aster yellows is a disease caused by a phytoplasm.

Nematodes are disease agents that are often overlooked but they can cause considerable yield losses. The adult nematodes have limited active movement but their eggs contained within cysts (dead females) are readily spread with soil, water, equipment or seedlings.

Parasitic nematode species that cause damage on napa cabbage:

- Heterodera schachtii
- Meloidogyne hapla
- Nacobbus batatiformis
- Rotylenchulus reniformis.

Cruciferous Vegetables

Broadly, cruciferous vegetables belong to the Cruciferae family, which mostly contains the Brassica genus, but does include a few other genuses. In general, cruciferous vegetables are cool weather vegetables and have flowers that have four petals so that they resemble a cross.

In most cases, the leaves or flower buds of cruciferous vegetables are eaten, but there are a few where either the roots or seeds are also eaten.

List of Cruciferous Vegetables

Extensive selective breeding has produced a large variety of cultivars, especially within the genus *Brassica*. One description of genetic factors involved in the breeding of *Brassica* species is the Triangle of U.

The taxonomy of common cruciferous vegetables			
common name	genus	specific epithet	Cultivar group
Horseradish	*Armoracia*	*rusticana*	
Land cress	*Barbarea*	*verna*	
Ethiopian mustard	*Brassica*	*carinata*	
Kale	*Brassica*	*oleracea*	Acephala group
Collard greens	*Brassica*	*oleracea*	Acephala group
Chinese broccoli (gai-lan / jie lan)	*Brassica*	*oleracea*	Alboglabra group
Cabbage	*Brassica*	*oleracea*	Capitata group
Savoy cabbage	*Brassica*	*oleracea*	Savoy Cabbage group
Brussels sprouts	*Brassica*	*oleracea*	Gemmifera group
Kohlrabi	*Brassica*	*oleracea*	Gongylodes group
Broccoli	*Brassica*	*oleracea*	Italica group
Broccoflower	*Brassica*	*oleracea*	Italica group × Botrytis group
Broccoli romanesco	*Brassica*	*oleracea*	Botrytis group / Italica group
Cauliflower	*Brassica*	*oleracea*	Botrytis group
Wild broccoli	*Brassica*	*oleracea*	Oleracea group

Bok choy	*Brassica*	*rapa*	*chinensis*
Komatsuna	*Brassica*	*rapa*	pervidis or komatsuna
Mizuna	*Brassica*	*rapa*	*nipposinica*
Rapini (broccoli rabe)	*Brassica*	*rapa*	*parachinensis*
Choy sum (Flowering cabbage)	*Brassica*	*rapa*	*parachinensis*
Chinese cabbage, napa cabbage	*Brassica*	*rapa*	*pekinensis*
Turnip root; greens	*Brassica*	*rapa*	*rapifera*
Rutabaga (swede)	*Brassica*	*napus*	*napobrassica*
Siberian kale	*Brassica*	*napus*	*pabularia*
Canola/rapeseed	*Brassica*	*rapa/napus*	*oleifera*
Wrapped heart mustard cabbage	*Brassica*	*juncea*	*rugosa*
Mustard seeds, brown; greens	*Brassica*	*juncea*	
White mustard seeds	*Brassica* (or *Sinapis*)	*hirta*	
Black mustard seeds	*Brassica*	*nigra*	
Tatsoi	*Brassica*	*rosularis*	
Wild arugula	*Diplotaxis*	*tenuifolia*	
Arugula (rocket)	*Eruca*	*vesicaria*	
Field pepperweed	*Lepidium*	*campestre*	
Maca	*Lepidium*	*meyenii*	
Garden cress	*Lepidium*	*sativum*	
Watercress	*Nasturtium*	*officinale*	
Radish	*Raphanus*	*sativus*	
Daikon	*Raphanus*	*sativus*	*longipinnatus*
Wasabi	*Wasabia*	*japonica*	

Cauliflower

Cauliflower is one of popular vegetable and known as "Ghobi or Gobi" in India and this flower belongs to "Cruciferaceae" family often overshadowed by its green cousin broccoli.

The edible portion of the cauliflower is called 'Curd' surrounded by leaves narrower than those of cabbage.

Site Selection

Cauliflower prefers a very fertile, moist but well-drained soil, high in organic matter and with a pH of 6.0–7.5. Poor soil results in reduced-quality crops. Sandy soils are acceptable but may require more frequent watering. For this reason, a soil with good water-holding capacity is best. Irrigate regularly, as a consistent supply of moisture is critical; water stress during curd development can cause unmarketable heads.

Plant in a location that receives full sun. Supply adequate levels of nitrogen to keep the plants productive over a long season. If your soil is not high in fertility, side dressing may be needed.

Timing and Succession Planting

Cauliflower is a cool-loving crop and performs best at temperatures below 80° F/27°C. Occasional temperature spikes generally do not damage the crop extensively but may result in ricey curds. Persistently hot weather often results in crop failure or reduced quality heads.

Varieties have been bred to succeed in specific harvest slots; for example, some varieties are better adapted to warm temperatures. Plan sowing and transplanting dates to ensure you are choosing the appropriate variety for the season.

For summer harvests: Select varieties that are adapted to maturing in the warmer temperatures of summer. Start seeds in early spring, March–April, and transplant as soon as temperatures have moderated. Do not transplant until after the last frost, as cauliflower seedlings have less tolerance to cold than older plants.

For fall harvests: Fall harvests can be achieved in any location, regardless of climate. Start seeds in June–July, depending on your location, and transplant approximately 4 weeks later. In short season, northern areas, where the harvest window is shorter, seed early through mid-June.

For winter harvests: Winter harvests are successful in areas where winters are mild, and temperatures rarely go below 32° F/0°C. Start seeds in late summer and transplant September– February for harvest January–April, depending on variety. Cauliflower should be 60–75% of their full mature size prior to entering winter; plants are generally more cold hardy when not full grown. Growth will resume in the spring.

Transplanting

Four to six weeks before transplanting, sow 2–3 seeds per cell in 72-cell plug flats or 3–4 seeds per inch in 20-row flats, ¼–½ inch deep. Keep soil temperature over 70° F/21°C until germination and 60° F/16°C thereafter. A seedling heat mat can aid in maintaining the correct temperature during germination. Thin to 1 plant per cell after germination when plants have their first set of true leaves.

Ensure good air circulation and light. If you need to sow during the heat of the summer, using a shade cloth can help moderate temperatures in the greenhouse.

When seedlings are 4–6 inches tall, and no more than 4–5 weeks old. Older transplants are typically stressed and do not perform as well as younger, actively growing seedlings. Prior to transplanting, gradually introduce the seedlings to increasing cold to harden. Transplant outdoors 18 inches apart in rows 24–36 inches apart. If there is disease pressure in your area, a wider spacing can promote better air circulation.

To grow mini heads of cauliflower, decrease the plant spacing. Plant 12 inches apart in rows 18–36 inches apart. Heads should be harvested at the desired market size.

Direct Seeding

Transplanting is the recommended, and most effective, method of planting cauliflower, but direct seeding is possible. Sow seeds ½ inch deep, 3 seeds every 18 inches in rows 24–36 inches apart. Thin to 1 plant every 18 inches when the first true leaves have formed.

Mini Cauliflower Heads

Head Formation

The best crops of cauliflower are grown during mild, cool weather. Crops can be grown successfully in warm temperatures, but extremely hot weather produces unmarketable heads. Select varieties based on local weather conditions. You can also consult your local Cooperative Extension Service agency for general guidelines on planting times for your area.

Excess cold weather can also be problematic. If young seedlings are repeatedly exposed to cool temperatures below 50° F/10°C for more than 7 days, plants may develop heads prematurely. This physiological issue is also known as "buttoning". Time seeding and transplanting to avoid excessive exposure to temperatures below 50° F/10°C. Buttoning can also be cause by nitrogen deficiency and inadequate irrigation. Due to their small size, buttoned heads are generally unmarketable.

Blanching

Other times referred to as "tying", many varieties can benefit from blanching the heads, especially during higher temperatures. Heads maturing in fall typically have less yellowing than those maturing in warmer weather. For white varieties, exposure to sun can cause the heads to yellow. When the heads are the size of a baseball, gather the outer, wrapper leaves and pull them over the heads, and secure with twine or a rubber band. Avoid securing the leaves too tightly over the heads, by attaching the twine or rubber band only at the end of the leaves — this allows the head ample room to continue development and increases air circulation.

Another method of blanching is to crack the midribs of the leaves and fold them over the head until it is fully covered. Do not break the leaves completely. Some varieties are described as "self-wrapping". These have wrapper leaves that cover the heads naturally and may not need manual intervention. However, self-wrapping varieties can also benefit from blanching the heads. Observe your crop and use your best judgment.

Tied or covered heads may experience increased humidity and a greater likelihood of contracting Alternaria.

Blanching Cauliflower

Diseases

To control diseases, adhere to a strict preventative program that includes long crop rotations (of at least 3 years) with non-Brassica crops, use clean starting mixes when sowing, and follow good sanitation practices. Should disease occur in your crop, have an infected specimen tested to positively identify the disease, and contact your local Cooperative Extension Service for potential control methods.

Pests

Insect pests common to any Brassica crop also affect cauliflower: aphids, flea beetles, and cabbage worms. Prevent the occurrence of pests by plowing in or removing debris of previous Brassica plantings, and practicing crop rotation. Exclude pests, such as flea beetles and cabbage worms by installing fabric row covers immediately after transplanting. Should flea beetle populations cause heavy pressure, treat with pyrethrin or azadirachtin.

Harvest

When the heads are at least 5–6 inches across, harvest by cutting at the base of the head. Waiting too long to harvest can cause the heads to be oversized and loose or ricey. Take care when handling cauliflower heads, as they are readily susceptible to bruising. Cool immediately after harvest.

Storage

Ideal storage conditions are in a cool location, 32° F/0°, with a relative humidity of 95–98% and good air circulation. Under these conditions, heads may remain good for 2–3 weeks. Warmer storage temperatures will decrease storage length.

Cabbage

The cabbage is a popular cultivar of the species Brassica Family and is used as a leafy green vegetable.

The only part of the plant that is normally eaten is the leafy head; more precisely, the spherical cluster of immature leaves, excluding the partially unfolded outer leaves. Cabbage is used in a variety of dishes for its naturally spicy flavour. The socalled "cabbage head" is widely consumed raw, cooked, or preserved in a great variety of dishes. Cabbage is an excellent source of vitamin C. It also contains significant amounts of glutamine, an amino acid that has antiinflammatory properties.

Climatic Conditions for Cabbage Farming

Cabbage grows best in cool moist climate and is very hardy to frost. In areas with comparatively dry atmospheres, its leaves tend to be more distinctly petiole than in the more humid areas. In hot dry atmosphere, its quality becomes poor and much of its delicate flavor is lost. Its germination is best at a soil temperature of about 55° F to 60° F. Temperatures below this and above this are not suited for it. Well hardened seedlings can tolerate temperature of 20° F to 25° F. It is grown mainly as rabi crop during winter. But in and around Nasik (Maharashtra), Ootacamond (Madras), and in semi parts of Kerala, it is grown as kharif crop also.

Best Soil for Cabbage Farming

Cabbage is grown in varied types of soils ranging from sandy loam to clay. It requires a pH ranging from 5.5 to 6.5 for higher production. It also thrives best when the soil is rich in organic matter and good drainage. Land/Field Preparation in Cabbage Farming : Land is prepared by ploughing it 3 to 4 times. The first ploughing should be done by soil turning plough, and the bulky organic manures should be spread in the field. Then the land should be ploughing and levelling the land, beds of suitable size and irrigation channels are made.

Seed Rate and Time of Sowing in Cabbage Farming

Normally 120 grams of seeds are required for one acre. Apply 480 Kgs of dry manure into a seedling bed of 160 m², and then sow the seed on the seedbed. This should produce sufficient seedlings for one acre of field. Cabbage is grown mainly as Rabi crop during winter (Sept.Oct.), But around Nasik (Maharashtra) it is grown as kharif crop also.

Transplanting and Spacing

Transplant the seedlings at 4 5 true leaves stage, about 25 days after sowing. Usually space them 45 cm apart in double rows of 4560 cm apart on each bed of 90 100 cm wide. Cabbage plant Spacing:

- Early maturity – Row to Row : 45 cm, Plant to Plant : 30 cm
- Late Maturity – Row to Row : 60 cm, Plant to Plant : 45 cm

Irrigation/Water Supply

Provide continuous supply of moisture. Install drip system with main and submain and place the inline laterals at the interval of 1.5. Place the drippers at the interval of 60 cm for 4 LPH or 50 cm for 3.5 LPH in the lateral system. Form the raised beds at 120 cm width at an interval of 30cm and place the laterals at the centre of each bed. Interval between two irrigations depends upon climate, soil and plant growth. In winter season irrigation at an interval of 810 days is sufficient.Cabbage cannot tolerant drought. Therefore irrigation should be applied frequently and evenly, especially in the head developing period. Irrigation should be applied following the first and the second side dressing. It is better to keep a little water in the furrow in the hot season. But drainage must be carried out in the rainy days.

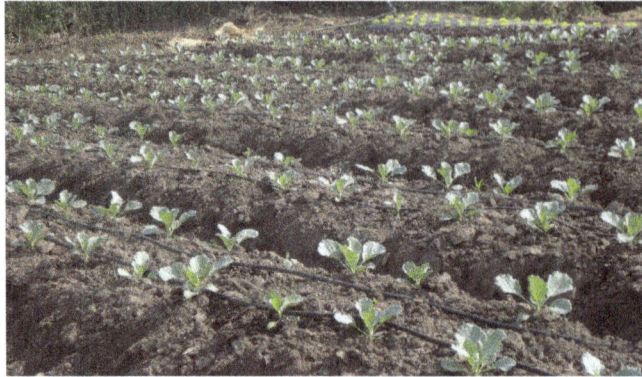
Cabbage Farming with Drip Irrigation

Pruning and Weed Control

It is necessary to remove the side shoots as soon as possible.Weeds must be removed as early as possible by hoeing but not too deep to damage the roots. Hoeing should not be done during the latter part of the growing season. Herbicide can be used for weed control in the cabbage field.

Fertilizers

It is better to use urea instead of Ammonium Sulphate where the soil is relatively acidic.If the soil is boron deficient, 5 –10 kg/ha borax should be applied before land preparation.For basal fertilizer, manure should be applied into the rows before chemical fertilizer.Chemical Fertilizers: Fertilizer application varies with soil fertility.Basal application before transplanting: 25:50:60 NPK kg / acre. First top dressing 1015 days after transplanting: 25:50:60 NPK kg / acre. Second application 20 – 25 days after first top dressing: 25:00:00 NPK kg / acre.Third application 1015 days after second application: 25:00:00 NPK kg / acre. Boron & Molybdenum should be sprayed at button stage.

Harvesting in Cabbage Farming

Cabbage is normally harvested when the heads reach full size and are firm,In cabbage harvesting is done depending on the maturity of the head and demand in market. Normally harvesting is done when heads are firm. If prices are high in the market harvesting is done earlier when heads are small and loose Heads are cut with a knife with little stalk arid some leaves. Proper grading is followed before heads are sent to market.

Savoy Cabbage

Savoy cabbage belongs in the *Brassica* genus along with broccoli and Brussels sprouts. This low calorie veggie is used both fresh and cooked and is high in potassium and other minerals and vitamins A, K and C.

The most obvious difference between common green cabbage and savoy is its appearance. It has multi-hued shades of green foliage that is typically tighter at the center, gradually unfurling to reveal curly, puckered leaves. The center of the cabbage looks a bit brain-like with raised veins running throughout.

Although the leaves look like they might be tough, the wonderful appeal of savoy leaves is that they are remarkably tender even when raw. This makes them perfect for use in fresh salads, as vegetable wraps or as a bed for fish, rice and other entrees. And they make even tastier coleslaw than their green cousin. The leaves are milder and sweeter than those of green cabbage.

Growing Savoy Cabbage

Growing savoy cabbage is similar to growing any other cabbage. Both are cold hardy, but savoy is by far the coldest hardy of the cabbages. Plan to set out new plants in the spring early enough so they can mature before the heat of summer. Sow seeds 4 weeks before the last frost for plants to be transplanted in June and plant fall cabbage 6-8 weeks before the first frost of your area.

Allow the plants to harden and acclimate to the colder temps before transplanting. Transplant the savoy, allowing 2 feet between rows and 15-18 inches between plants in a site with at least 6 hours of sun.

The soil should have a pH of between 6.5 and 6.8, be moist, well-draining and rich in organic matter for the most optimal conditions when growing savoy cabbage.

If you start with these requirements, caring for savoy cabbage is fairly labor free. When caring for savoy cabbage, it's a good idea to mulch with compost, finely ground leaves or bark to keep the soil cool, moist and retard leaves.

Keep the plants consistently moist so they don't stress out; apply 1- 1 ½ inches of water per week depending upon rainfall.

Fertilize the plants with a liquid fertilizer, such as fish emulsion, or 20-20-20 once they develop new leaves, and again when the heads begin to form.

Kale

Kale (Brassica oleraceae var acephala), commonly known as Sukumawiki, is a cool season crop that belongs to the Brassicas family. Its leaves, which are rich in vitamins and essential mineral elements are widely utilized alone or mixed with other vegetables, pulses or meat.

It is an all-year crop and has the potential to reduce poverty levels.

The demand for kales is usually very high because of their benefits, which include the following;

- High in iron which is essential for good health, e.g. in formation of haemoglobin and enzymes, cell growth, proper functioning of the liver, among others.

- Rich in vitamin K which helps in protecting the body against various cancers and is also necessary for a wide range of bodily functions.

- High in fibre, low in calorie and zero fat.

- Has anti-inflammatory properties which helps in fighting against asthma, arthritis and autoimmune disorders.

- Rich in antioxidants, e.g. flavonoids and carotenoids which protect the body against various cancers.

- Rich in vitamin A and C.

- Good source of calcium which helps in preventing osteoporosis, bone loss and maintaining a healthy metabolism.

- Rich in zeaxanthin and lutein which are powerful nutrients which protect the eyes.

Kale is grown in a wide range of climatic conditions provided water is available.

- Soils- performs best in well drained soils which are rich in organic matter with a pH of 5.5- 7.5.

- Altitude- does well at altitudes of 800-2200M above sea level.

- Temperature- requires an optimum temperature range of 16-21°.

- Rainfall- the crop requires sucient amounts of moisture throughout the season. A well distributed rainfall of 30-500mm is ideal for optimum yield. Irrigation is recommended if rainfall is inadequate.

- Kale can tolerate slightly alkaline soil and frost.

- It requires at least 6 hours of direct sunlight daily and prefers plentiful, consistent moisture.

- Some varieties are drought tolerant.

Kohlrabi

Kohlrabi is a cool season vegetable that prefers a sunny location and fertile, well-drained soil. Incorporate plenty of organic matter and a complete fertilizer into the area before planting. Plant seeds ¼-¾ inch deep, 1-2 weeks before the last frost in the spring. Thin seedlings or transplant kohlrabi 6 inches apart in the row with rows 1 foot apart. Irrigate regularly, and avoid water or fertilizer stress during growth. Kohlrabi is the least hardy of the cabbage like vegetables. Temperatures below 45° F will cause the plant to flower. Hot weather causes the stem to become woody and tough. Control insects and diseases throughout the year. Harvest kohlrabi when the stem enlarges to 2-3 inches in diameter.

Recommended Varieties

There are many good kohlrabi varieties for sale in local gardening outlets and through seed catalogs. Most grow well in Utah. Rapid (45 days, purple skin), Grand Duke (50 days), White Vienna, and Purple Vienna (60 days) have excellent production, eating quality, and flavor.

Methods to Grow Kohlrabi

- Soils: Kohlrabi prefers fertile, well-drained soil rich in organic matter for best growth. Most soils in Utah are suitable for kohlrabi production.

- Soil Preparation: Before planting, incorporate 2-4 inches of well-composted organic matter and apply 4-6 cups of all-purpose fertilizer (16-16-8 or 10-10-10) per 100 square feet.

- Plants: Kohlrabi can be grown from seed or transplants. Seeds should be planted ¼-¾ inch deep and thinned to the final stand when plants have 3-4 true leaves. Plants removed at thinning can be transplanted to adjacent areas. Transplants can be used to provide earlier harvest. Transplants should have 4-6 mature leaves and a well developed root system before planting. Generally 4-6 weeks are required to grow transplants to this size.

- Planting and Spacing: Seeded or transplanted kohlrabi should be spaced 6 inches between plants in the row with rows 1 foot apart. Kohlrabi grows best when temperatures do not exceed 75° F. Young plants may be damaged by hard frosts. Mature plants will flower if average temperatures during growth are less than 45° F. Transplants may be planted 1-2 weeks before the last frost date for the growing area. Seeded kohlrabi may be planted at the same time. For fall maturing kohlrabi, select early maturing cultivars and plant 50 days before the anticipated maturity date. The maturity date can be timed for 2-3 weeks after the first fall frost. High summer temperatures reduce growth, decrease quality, and cause the enlarging stems to become tough and woody. In hot areas it is best to grow kohlrabi as a spring or autumn crop.

- Water: Water kohlrabi frequently, since roots are shallow. About 1-2 inches of water are required per week. Use drip irrigation if possible to conserve water. Applying mulch around the plant also helps conserve soil moisture and reduces weed growth. Moisture fluctuations will cause the stems to become tough and woody.

- Fertilization: Apply 1 cup per 10 feet of row of a nitrogen-based fertilizer (21-0-0) 3 weeks after transplanting or thinning to encourage rapid plant growth. Place the fertilizer 6 inches to the side of the plant and irrigate it into the soil.

- Mulches and Row Covers: Plastic mulches can help conserve water, reduce weeding and allow earlier planting and maturity, especially with transplants. Fabric covers are used to protect seedlings and transplants from frosts and insect pests. Apply organic mulches when summer temperatures increase. These will cool the soil and reduce water stress. Organic mulches such as grass clippings, straw, and newspaper also help control weeds.

Problems

- Weeds: Plastic and organic mulches effectively control weeds. Be sure to control weeds when plants are small and be careful not to damage roots when cultivating.

Insects and Disease

Insect	Identification	Control
Aphids	Green or black soft-bodied insects that feed on underside of leaves. Leaves become crinkled and curled.	Use insecticidal soaps, appropriate insecticides, or strong water stream to dislodge insects.
Cabbage Worms and Loopers	Worms and loopers are light to dark green. Adult loopers are gray or brown moths while cabbage worms are white butterflies. Worms and loopers chew holes in leaves and hide in kohlrabi leaves.	Control worms and loopers with appropriate insecticides or biological measures.

Flea Beetles	Small black beetles that feed on seedlings. Adults chew tiny holes in cotyledons and leaves. Beetles can reduce plant stands or may kill seedlings.	Control beetles with appropriate insecticides at seeding or after seedlings have emerged from the soil.

Harvest and Storage

Kohlrabi should be harvested when the stems reach 2-3 inches in diameter. Larger stems tend to be tough and woody. The young leaves can also be eaten like cabbage or kale. Kohlrabi can be stored for 2-3 weeks at 32° F and 95% relative humidity. When prepared, the outer skin is peeled off and the inner flesh is eaten raw or cooked. Kohlrabi tastes like turnips with a texture like water chestnuts.

Productivity

Plant 3-5 feet of row per person for fresh use and an additional 5-10 feet of row for storage or processing purposes. Expect 50-75 lbs per 100 feet of row.

Broccoli

Broccoli is a hardy vegetable that develops best during cool seasons of the year. Two crops per year (spring and fall) are possible in most parts of the country, especially with continuous improvement in fast maturity and heat tolerance that extends the life of broccoli through all but the hottest parts of the season. It belongs to the cole crop family (Brassica oleracea), which includes cabbage, Brussels sprouts, cauliflower, collards, kale, and kohlrabi.

Site Selection

Broccoli grows best on well-drained soils that have good water-holding characteristics. If you grow broccoli on sandy soil, irrigation is important for optimum plant growth and proper main head and side shoot development.

Planting and Fertilization

Because broccoli is a cool-season crop, it generally is planted in the spring. You should begin planting when soil temperatures reach at least 50° F and the possibility of hard frosts (28° F or lower) has passed in your area. Flower heads (the edible portion of broccoli) develop relative to ambient temperatures, and in the heat of summer, broccoli heads maturing in July may bolt (produce flowers and seeds) more quickly (4-6 days) than those maturing in the cooler spring and fall periods. Broccoli heads must be closed and tight (no yellow petals showing) to be considered good quality.

While broccoli generally is transplanted in the spring, it can be sown directly from seed in late summer or early fall, when soil temperatures are in the high 60s and ambient air temperatures are in the 80s. Under these conditions, seeds generally emerge in less than 7 days. Adequate soil moisture is essential for optimum broccoli seed germination. Depending on climate, transplanting begins in late March to mid- April. Successive plantings can occur every 2 weeks through August.

Optimal plant populations for broccoli are 14,000 to 24,000 plants per acre. Therefore, the amount of seed per acre that you should buy varies with plant spacing, final plant stand, and percent seed germination. Depending on the planter type used (random or precision), you should sow 0.5-1.5 pounds of broccoli seed per acre, with seeds placed 12-18 inches apart in 36-inch rows. When transplanting, you should have a minimum of 11,000 plants per acre. Spacing decisions depend on the row spacing of your equipment, your ability to irrigate, the planting date, and your specific market requirements (small or large heads).

Fertilizer rates should be based on annual soil test results. If you are unable to conduct a test, the recommended N-P-K application rates are 120-100- 100 pounds per acre broadcast or 35-50-50 pounds per acre banded at planting. Liming may also be necessary to maintain soil pH in the 5.8-6.6 range for optimal growth. Cruciferous crops such as broccoli require more boron than most other crops. Applying 3 pounds of boron per acre will eliminate broccoli stems that are brown and hollow. Severe boron deficiency can produce browning on head surfaces. These affected heads are not marketable.

Pest Control

Weed control can be achieved with herbicides, mechanical control, and a good crop-rotation system. Broccoli competes fairly well with weeds, but it should be kept weed-free until plants reach the preheading stage. Many pretransplant and postemergence herbicides are available for broccoli, depending on the specific weed problem and the broccoli growth stage. If infestation levels are mild, cultivation can be used to reduce weed problems.

Insects are a major potential problem in broccoli production. Flea beetles, cabbage loopers, imported cabbageworms, diamondback moths, and aphids all can cause crop losses. Monitoring insect populations with traps or by scouting will help you determine when you should use pesticides and how often you should spray.

Several broccoli diseases can cause crop losses. Black rot, blackleg, bacterial head rot, downy mildew, and *Alternaria* are common problems. Many of these diseases can be prevented by having a good crop-rotation program and by using disease-resistant varieties.

Harvest and Storage

Because there are no mechanical harvesters for broccoli, it is necessary to hand-harvest the crop. To ensure marketing a high-quality product, you should check the broccoli heads for worms, which tend to hide underneath the florets. You also will need to grade the heads for size (head diameter generally averages 6 inches) and for flower bead tightness.

Broccoli should be cooled with packed ice or a hydrocooler immediately after harvest. Broccoli that is cooled and maintained at 32° F and 95-100 percent relative humidity can be stored for 10-14 days. If broccoli is stored this long, however, it will begin to lose its dark green color and firmness.

Brussels Sprout

Brussels sprouts are a classic cold-hardy fall vegetable, recognized by the small sprouts lining the

2–3 foot tall stalk. A close relative to cabbages, each sprout looks like a miniature cabbage. They have a long maturity period, typically 90–120 days, and often are best harvested after the first few frosts as cold weather improves their flavor.

Site Selection/Crop Care

Brussels sprouts will produce the best crops when planted in full sun, though they will tolerate part shade, in average soils with moderate fertility and a pH above 6.0.

To maintain steady growth, the crop should be well fertilized and irrigated in dry weather; plants should be provided the equivalent of 1 inch of water per week. Crops grown under poor fertility or dry conditions will yield sprouts of lesser quality. Cool weather will provide the best growing conditions, especially when forming sprouts, but the plants will grow well in areas with mild summer weather prior to sprout formation.

Fertilization

Precisely timed fertilization has a direct correlation with the quality of the sprouts produced. Plants should not be fertilized after early July. Stopping fertilization initiates sprout formation and ceases growth of the stem. At this time, the leaves may begin to yellow and drop off the stem; this is normal.

Fertilization techniques can vary depending on the size of the area planted, the capabilities of your equipment, and how much fertilizer your crop needs.

For Direct-to-market Production

Prior to planting, perform a soil test to identify your fertilization needs. For most varieties, approximately 200 pounds of nitrogen per acre is required for a good crop. Fields that already have a sufficient amount of nitrogen in the soil, as indicated by the soil test, won't require side dressing unless the plants appear to be growin gslowly from lack of nitrogen. If your soil test indicates a nitrogen deficiency of 50 pounds or less we recommend side dressing that amount of nitrogen 4–6 weeks after transplanting. If the deficit is greater than 50 pounds we recommend side dressing twice, using half the quantity of the deficit each time, 3 weeks and 7 weeks after transplanting. Note that these recommendations are only guidelines and that you may want to experiment with them. If your growing methods do not allow for side dressings then we recommend planting in soil with 200 pounds of nitrogen per acre.

Transplanting

While direct seeding is possible, it is recommended to transplant Brussels sprouts. Sow 2–3 seeds per cell in 72-cell plug flats or sow 3–4 seeds per inch in 20-row flats, ¼ inch deep, 4–6 weeks before transplanting. Keep temperatures at 75° F/24°C until germination occurs. When seedlings are 3–4 inches tall, transplant with 18 inch spacing between plants, in rows 24 inches apart.

Diseases

To control diseases, adhere strictly to a preventative program that includes long crop rotations, at least 3 years, with non-cruciferous crops, clean starting mixes, and strict sanitation practices. Should disease occur in your crop, have an infected specimen tested to positively identify the disease.

A common disease of Brussels sprouts is black rot, identified by yellow lesions on the leaves in its earliest stages. As the disease progresses, the affected leaves may die and turn brown to black. Johnny's only offers seed that has been tested free of black rot in a sample of 30,000 seeds.

There are two host-specific species of Alternaria mold that affect Brussels sprouts and other Brassica. Small, dark spots that later expand into larger, tan circles are present on infected leaves. Alternaria favors wet conditions, so ensure proper air ventilation to prevent it. Refer to the disease control chart in our catalog or on our website for fungicides applicable to the treatment of Alternaria.

Pests

The best insect pest control on young plants is the use of floating row covers, which prevents the insects' access to the plants. Put row covers in place on the day of transplanting. If heavy pressure from flea beetles is observed, treat with azadirachtin or pyrethrin. Cabbage worms can be controlled with Bacillus thuringiensis (B.t.). The presence of cutworms can be prevented by cultivating the soil 2–4 weeks before transplanting seedlings to work in any cover crops and destroy weeds.

One of the most prevalent insect pests in Brussels sprouts are cabbage aphids (Brevicoryne brassicae). Aphids occur in dense groups and can be identified by their white, waxy appearance. They tend to be attracted to the young leaves that form the sprouts, causing a problem in the harvested crop.

Harvest

Prior to harvest it is common for the leaves of the plant to senesce and turn yellow. It is the result of the plant pulling nutrients from the leaves and directing them to the developing sprouts, not the effects of a nutrient deficiency.

Sprouts can be harvested after the first frost and until the end of December in most areas, and through the winter in areas where the cold is not severe. Pick sprouts when they are firm and well formed, generally beginning when the lower leaves start to turn yellow. Break off the leaf below the sprout and snap off the sprout.

If the plant was not topped, the upper sprouts will continue to form and enlarge as the lower ones are harvested. If the plants were topped, the entire stalk may be harvested at once by cutting the stem below the lowest sprouts.

Storage

Store in a cooler or cold cellar at 36° F/2°C with 95–98% relative humidity. They will store for 4–6 weeks under these conditions.

References

- Millard, E. (2014). Indoor Kitchen Gardening: Turn Your Home Into a Year-round Vegetable Garden. Cool Springs Press. p. 63. ISBN 978-1-61058-981-9. Retrieved May 28, 2017

- Easy-guidelines-spinach-swiss-chard: africanfarming.com, Retrieved 25 May 2018

- Gupta P, Kim B, Kim SH, Srivastava SK (Aug 2014). "Molecular targets of isothiocyanates in cancer: recent advances". Molecular Nutrition & Food Research. 58 (8): 1685–707. doi:10.1002/mnfr.201300684. PMC 4122603. PMID 24510468

- Chinese-cabbage-farming: agrifarming.in, Retrieved 20 March 2018

- Barclay, Eliza. "Introducing Microgreens: Younger, And Maybe More Nutritious, Vegetables". NPR. Retrieved 23 January 2014

- Cabbage-farming: agrifarming.in, Retrieved 21 March 2018

- Xiao, Z.; Lester, G. E.; Luo, Y.; Wang, Q. (2012). "Assessment of Vitamin and Carotenoid Concentrations of Emerging Food Products: Edible Microgreens". Journal of Agricultural and Food Chemistry. 60(31): 7644–7651. doi:10.1021/jf300459b. PMID 22812633

- Growing-savoy-cabbage, vegetables: gardeningknowhow.com, Retrieved 18 June 2018

- Afable, Patricia O. (2004). Japanese pioneers in the northern Philippine highlands: a centennial tribute, 1903-2003. Filipino-Japanese Foundation of Northern Luzon, Inc. p. 116. ISBN 978-971-92973-0-7

- Growing-broccoli: bonnieplants.com, Retrieved 19 April 2018

- Bibi Z (2008). "Role of cytochrome P450 in drug interactions". Nutrition & Metabolism. 5: 27. doi:10.1186/1743-7075-5-27. PMC 2584094. PMID 18928560

3

Root Vegetables

Root vegetables are underground parts of a plant that are consumed by humans. These can be of various types, such as bulb, modified plant stem, root-like stem and true roots. The topics elaborated in this chapter address these different types of root vegetables.

Root vegetables are underground plant parts that comprise a substantial part of year round vegetable crops. Root vegetables are rich in flavor, economical, versatile and especially good from October to March when our bodies crave heartier fare. Consider the list of vegetables pulled from below ground; potatoes, garlic, carrots, radishes, onions, beets, sweet potatoes, yams, turnips and many more. Root vegetables are staples of winter crops and are used in all aspects of restaurant menus-salads, soups, entrees, stews and as substantial side dishes. While the more common roots-potatoes, carrots, beets-are used year round, it is during the colder months that the less glamorous roots appear-turnips, parsnips, celery root, rutabagas and sunchokes. These lesser known roots are often referred to as lowly vegetables, not so much because of their below ground location, but rather of their overall status in the vegetable kingdom. These roots have enjoyed a renaissance of sorts and are now more prevalent in winter recipes and menus than in recent years. In general, root vegetables are low in calories, contain virtually no fat and add fiber and vitamin C to daily diets. The more deeply colored roots-carrots and beets offer beta carotene (vitamin A) and antioxidants that contribute to good health.

Bulb Root Vegetables

Bulb vegetables have relatively large, usually globe-shaped, underground buds, or bulbs, with overlapping leaves arising from a short stem. Common bulb vegetables include onions and garlic. The many types of tart, pungent bulbs known as onions (Allium cepa) are among the world's oldest cultivated plants. Bulbs usually grow just below the surface of the ground and produce a fleshy, leafy shoot above ground. They usually consist of layers, or clustered segments.

Onion

The onion (*Allium cepa* L.), is the most common vegetable of the genus Allium, which includes several important vegetables: *Allium cepa* (shallot, top set onion, multiplier onion), *A. sativum* (garlic), *A. ampeloprasum* (elephant garlic or great head leek), *A. schoenoprasum* (chive), *A. fistulosum* (Welsh onion, Japanese bunching onion), *A. chinense* (rakkyo), *A. tuberosum* (Chinese chive), and *A. cepa* x *A. fistulosum* (Beltsville bunching onion). These monocots were once classified in the family Alliaceae but have been reclassified into the subfamily Allioideae of the family Amaryllidaceae. Onion, is a biennial or a perennial by means of its bulbs in areas mild enough to over winter. Alliums are cool season crops, tolerant of frost but not prolonged freezing tempera-

68 Vegetable Crops

tures below the range from -9.5 to -6.5°C. Optimum temperatures for growth and development are in the range from 13 to 25°C, somewhat narrow compared to other vegetables. Flowering occurs after vernalization .

Alliums have a unique plant architecture. The plant consists of a short subconical stem from which the linear leaves arise in 1/2 phyllotaxy (all leaves lie in a single plane). The leaves arise from the short crown stem in a compact series, and the sheaths of the outermost leaves enclose the younger ones. The sheathing base of each leaf completely encircles the stem, and it is the development of the fleshy stem bases that, together with the lack of internodal elongation, results in the development of the bulb. Alliums, except leeks and chives, are composed of fleshy, enlarged leaf bases or scales. Chives do not produce swollen bulbs. Onion has long hollow leafless stems that arise from the terminal bud and bear the inflorescence. Flower stalks may also arise from lateral buds. The terminal inflorescence develops from the ring-like apical meristem. Scapes, one or more, elongate from 30 to more than 100 cm above the leaves. The scape is a long, leafless flower stalk extending between the spathe and the last foliage leaf. At first, the scape is solid but, by differential growth, becomes thin walled and hollow. The onion scape has a characteristic spherical bulge along its length. The total number of developing scapes depends on the number of sprouted lateral buds.

A spherical umbel develops atop each scape and can range from 2 to 15 cm in diameter. Early in development, a spathe initially encloses the inflorescence. The umbel is an aggregate of generally 200-600 small individual flowers each less than 5 mm in length. Flowering in a given umbel may continue four or more weeks but individual flowers are fertile for only a week. Flowers are perfect, with six white petals, six stamens, a single style, and an ovary with two ovules. Bulbils are produced in the inflorescence of some cultivars in place of florets. Onion and certain other Alliums exhibit protandry, the shedding of pollen before the stigma is fully developed and receptive. Protandry promotes out-crossing by insect pollinators. It takes about two weeks or more for all the florets on an umbel to open completely. Most of the pollen is shed on the first and second day a floret is open and pollen viability declines quickly after opening. The stigma develops more slowly usually after pollen release typically on days 3 to 4 post anthesis and may remain receptive on day 6 or 7 after opening. The flowers are self-fertile so pollen from one floret can pollinate another on the same umbel. Where male-sterile lines are used for seed production, pollen must be moved between umbels for pollination.

The garlic umbel inflorescence develops at the top of a straight and solid scape, is subspherical, and usually contains only bulbils or a combination of bulbils and flowers, which rarely if ever set seed in commercial clones. The infrequently formed flowers are lavender and usually wither and abort. Recently, researchers in Germany, Japan and USA have produced viable seeds from certain garlic clones particularly those belonging to subsp. *ophioscorodon* of *A. sativum* .

Soil Preparation

Alliums can be grown in a wide range of soil types but prefer fertile, well-drained, non-crusting mineral soils that are high in organic matter and have good moisture retention. Heavy clay soils and light sandy soils should be avoided unless frequent irrigation is available. Onions grow best in the pH range of 5.3 to 6.5. Above pH 6.5, deficiencies of copper, manganese, and zinc may occur. Onions should be grown in the same field once every four to five years to help prevent the buildup of soil-borne disease, nematodes, and weeds. The soil must be finely tilled to depth of at least 18 cm using rotovators or disks and free of clods, stones, and other impediments, especially for direct seeding, to ensure uniform emergence and good stand establishment. Pelleted seeds improve singulation and spacing with precision belt or air seeders. Planting Allium seed crops on raised beds improves drainage and encourages salt accumulation away from the root zone.

Fertilization

Onions are a shallow-rooted, cool-season crop that responds well to fertilization. Generally, Alliums have a high requirement for micro-nutrients and moderate need for N-P-K. Alliums are moderately sensitive to sensitivity to salinity especially at the time of germination and in the small seedling stage but become more tolerant as they mature. Significant yield reductions occur when salinity is in the range of 4 to 5 mmhos/cm. Alliums have a shallow fibrous root system restricted to the top 30 cm of the soil, so precision fertilizer placement is essential for efficient use. Fertilization needs vary depending on many factors such as temperature, cultivar, spacing, soil type, available water etc. Fertilization should be in accordance with soil test results so that costly mineral inputs are not wasted and become pollutants. Tissue and soil analysis combined cropping history are important for determining N fertilization needs. Quick tests of leaf tissue provide rapid assessment of N availability during the growing season.

Nitrogen requirements vary depending on nitrogen-supplying capacity of the soil, irrigation efficiency, and the amount of N loss due to rainfall and other environmental factors. Alliums are generally sensitive to ammonia and salts but tolerant of low pH compared to other vegetables. Composted manure (11 to 22 t/ha) can supply nutrients and fulfill early season N requirements. With efficient irrigation, a total of 260 kg/ha of N should maximize yield on most mineral soils, with less needed on soil types with high residual N. However, needs may vary from 110 to 450 kg/ha depending on soil, cropping history and water received. Higher amounts may be justified in fields receiving significant rainfall or lower mineral holding capacity. Excessive applications of N may cause excessive vegetative growth and delay flowering. Excess nitrogen also promotes unwanted secondary growth.

Soils with bicarbonate extractable P greater than 30 ppm require a preplant application of no more than 56 kg/ha of P_2O_5, while soils testing at 10 ppm P or less may require as much as 224 kg/ha of P_2O_5. With adequate preplant application, in-season P additions are usually not required.

Soils exceeding 150 ppm ammonia-acetate-exchangeable K are unlikely to respond to additional fertilization. However, if soils test at less than 100 ppm, potassium additions of up to 170 kg/ha of K may be needed to ensure adequate fertility for seed crops.

Nitrogen fertilizer should be delivered in multiple applications throughout the season, with not more than 25 percent of the seasonal total applied in a single application. Some growers row band all P or P and K fertilizer in the bed before planting. Nitrogen, and K if needed, is applied in various combinations such as a pre-plant band application at seeding, a top dressing approximately 4 weeks into the season as bolting begins, or weekly if required by fertigation in conjunction with petiole analysis results.

Production Methods

Seed to Seed

For the seed-to-seed method plants are direct seeded, vernalized in the field, bolt, and produce seed. Seed-to-seed production is less expensive because labor is reduced since bulbs are not lifted or replanted. Seed-to-seed production fields are seeded, usually in late summer, allowing plants to develop beyond the juvenile phase so that vernalization can occur during the fall or winter. Onion seeds germinate slowly at a base temperature of 4°C and optimal germination occurs at approximately 24°C. Before planting it is important to ensure the minimum isolation distances are met from other *Allium cepa* seed fields. Seed-to-seed onion crops are often planted on raised beds 100 to 105 cm wide although other widths are possible. For example, single rows may be planted on smaller 75 cm wide raised beds. Plate, belt, or vacuum precision seeders can plant to a final stand at optimal spacing so thinning is not required. The soil should be finely tilled with small ped size and kept moist until emergence. Target seeding depths are 10 mm to 2.5 cm deep with deeper planting on lighter drier soils. For 75 cm beds, the target plant population is 45 to 60 plants per m of row. For double rows, planted on wider 100 cm beds, the target plant populations are 50 to 70 plants per m of row or about 30 seed stalks per m². Both single and double row seeding densities require 3.5 to 5.5 kg/ha of seed.

Seed-to-seed production causes many short-day cultivars to develop multiple scapes, while long-day cultivars tend to produce a single scape. Seed-to-seed yields can be as high or higher than bulb-to-seed yields. However, the quality of seed from the bulb-toseed method is better because of the additional selection and roguing of bulbs for off-color and off-type that occur before planting for seed production. However, excellent quality can be maintained in seed-to-seed production, provided stock seed is grown from bulbs that are rigorously rogued for off types.

Bulb to Seed

The bulb-to-seed method is widely used for seed production of open-pollinated cultivars as well as stock or parental seed production. This method allows bulb characteristics to be assessed and selected to maintain genetic purity, uniformity, and trueness-to-type. Bulb-to-seed is also used to vary the planting dates of parental lines for hybrid production to synch flower times. Bulb-to-seed requires planting of mature nondormant bulbs that have been vernalized to induce bolting. Because bulbs are established more quickly, planting can be delayed until fall. Vernalization of bulbs by cold temperature to induce bolting can occur in field as long as they are exposed to below 10°C for extended periods. However, prolonged exposure to potentially lethal temperatures between -7 and -9°C or below must be avoided. Lethal temperatures for onion depend on genotype, plant size, soil conditions and temperature duration. Generally, bulbs of long-day cultivars are harvested in the fall and stored at low temperatures for spring planting. The resulting seed crop is harvested in the late summer or early fall the following year. Short-day cultivars are grown during winter and bulbs are harvested in the spring. These are stored during summer and replanting during the fall with seed harvest occurring in late spring or early summer of the following year. When lifted bulbs are inspected, the shape, color, and size should be checked for trueness-to-type. At replanting, these characteristics can be checked again as well as early sprouting bulbs. Bulbs are generally planted in single rows on beds at a spacing of 5.0 cm or closer depending on bulb size. The bulbs are placed in a shallow furrow and covered with 2.5 to 3.0 cm of soil. Bulbs tend to produce multiple seed stocks compared to seed-to-seed production.

Pollinators

Even though onion flowers are perfect their male and female flower parts develop at different times so they are cross pollinated by insects. The flowers are visited by a range of insects, including bees, flies, and other insects, that collect pollen and nectar. The duration of anthesis of an individual umbel is roughly 4 weeks. Honey bees reluctantly visit onion flowers to collect both nectar and pollen, but only nectar foragers will visit both male-sterile and male-fertile lines in hybrid onion production. Onion flowers, and specifically their nectar, are one of least favorite of honey bees who prefer flying longer distances to visit other flowers types instead. The sugar concentration of the nectar may be increased by potassium fertilizer. However elevated potassium in nectar may be another reason why bees find onion nectar unattractive. Therefore, seed producers should not plant crops more attractive to bees near onion seed fields. If honeybees are used, supplemental hives with colony stocking rates as high as 30 hives per hectare or more have been suggested.

Flies have been used for pollination of onions particularly during plant breeding and producing small lots of seeds. Alternative pollinators such as native bees, *H. farinosus* for example, play a role in onion pollination and farmscaping may increase their populations. Managed alternative bees, such as alkali bees (*Nomia melanderi*), are increasingly used to pollinate onion seed fields.

Have insecticides have negative impacts on pollinator attraction and pollen/stigma interactions, with certain products dramatically reducing pollen germination and pollen tube growth. Decreased pollen germination was not associated with reduced seed set; however, reduced pollinator attraction was associated with lower seed set and seed quality, for one of the two female lines examined. Our results highlight the importance of pesticide effects on the pollination process. Overuse may lead to yield reductions through impacts on pollinator behavior and post pollination processes Overall, in hybrid onion seed production, moderation in insecticide use is advised when controlling onion thrips, *Thrips tabaci*, on commercial fields.

F-1 Hybrids

F-1 hybrid cultivars are more uniform and higher yielding than open pollinated onions and this has led to their adaptation. Hybrid onion seed production requires special considerations. The time from seeding to bulb planting to flowering differs among inbred lines crossed to make F-1 hybrids. To achieve simultaneous flowering or 'nicking' of male and female parental lines, planting in most cases will need to occur at different times. An alternative approach is to direct seed one parental line and plant bulbs of the other. Seed producers avoid planting bulbs of both parental lines. The ideal situation is to develop inbred lines for seed production that have similar maturities but unfortunately this doesn't always occur. The ideal situation is to have the male pollen parent flower shortly before, during, and after the female parent flowers. To achieve pollen production over an extended period, split plantings are sometimes employed. For success, seed companies must share information about flowering characteristics of hybrid parental lines with seed producers.

Many popular modern onion cultivars are F-1 hybrids produced using a male sterile inbred line as the female parent to ensure crossing by bees without costly and time-consuming emasculation and hand pollination. Bees tend to find male-fertile lines more attractive than male-sterile lines and have a tendency to move up and down rows instead of crossing between male-fertile and male-sterile lines, which may reduce pollination. Because only bees foraging for nectar will visit both male-fertile and male-sterile lines, colonies introduced to onion fields should have large numbers of adult bees and should not be fitted with pollen traps or be fed with sugar syrup, as both methods promote pollen collection instead of nectar foraging.

Isolation Receipt

The minimum recommended isolation distance between different cultivars is 2000 m. Some authorities stipulate shorter distances than this for cultivars with the same bulb color. In some countries, there are declared zones in which only cultivars of a specific bulb color can be grown for seed.

Irrigation

After seeding, the soil should remain moist so germination and early seedling development are uniform. Moist soil also prevents crusting which may delay or inhibit seedling emergence, reduce uniformity, and potentially delaying maturity. Since seeding occurs in the summer, often at high temperatures, sprinkler irrigation is used to cool the soil and prevent crusting by evenly hydrating the seed bed. Since onions have a shallow root system that is composed of straight non-branched roots that extend from the basal stem plate, plants must receive a consistent supply of water throughout the growing season. Onions cannot access water at soil depths below 60 cm, and most available water is extracted from the top 30 cm of soil. The general recommendation for at least one inch of water per week, especially during bulbing, applies to onions. Photosynthesis and expansive growth are reduced by even mild water stress, because unlike many crops, onions cannot reduce their leaf water potential by osmotic adjustment to compensate for dry soils. Stressed onion plants exhibit poor flower and pollen development, reduced seed yields, lower seed weights, reduced nectar production and lower seed vigor when germinated. The amount and frequency of irrigation required depends on soil type, stage of crop development, environmental conditions (humidity, wind, irrigation methods, rainfall, evapotranspiration etc.). Evapotranspiration from a field can be monitored by pan evaporation with tensiometer measurements of soil to determine when water application is needed. Crop coefficients published for various locations determine how much of the evaporative losses must be reapplied when the soil moisture tension threshold is reached. Irrigation is usually needed when 25% of available water in the top 60 cm has been depleted. In general, onion seed crops use 65 to 90 cm of water per growing season. With 70 to 80% efficiency water application of 90 to 115 cm may be required to produce a seed crop.

Growth and Development

Onion food crops are grown as an annual for fresh use when immature (green onions) or mature (dried bulbs). However, for seed production, onions require two seasons to complete their reproductive growth cycle. The initial growth rate is slower than many other cool-season crops because of slow leaf area development and poor light interception that limits photosynthesis. As seedlings become established and grow, new foliage and roots continue to be produced, along with a slight elongation and widening of the compressed stem. Onion seed production requires low-humidity conditions during spring and summer as seeds are maturing. Disease management, pollination, and seed maturation are enhanced by warm temperatures and low relative humidity. Foliage diseases are more common under humid conditions and bee pollination is reduced during rainy weather. Pre- or post-harvest drying is enhanced in climates with low humidity. Climates that are cool in the winter with warm to hot spring and summer seasons with low humidity and limited rainfall are near ideal for onion seed production. Although onion seed production requires two growing seasons, both can occur within a single 10- to 12-month window if the crop is seeded in late summer. The first season the plants grow vegetatively from seed and produce a bulb based in response to day length. The optimum range of temperatures for best growth and development are 20 to 25°C.

Bulbing

Bulbing is a change in leaf morphology initiated when sufficient exposure to a critical day length is exceeded, although temperature has an influence as well. Each cultivar has a critical day length for bulbing induction. The duration of light exposure is most important, and the exposure process is cumulative. A brief exposure to the appropriate day length stimulus is not sufficient to initiate the bulbing process. When cultivars reach their critical day length before adequate vegetative growth is achieved, resulting bulbs will be small. Cultivars that require long days to bulb will not bulb when grown during short days.

Onions are identified as short-, intermediate-, or long-day cultivars. Short-day cultivars bulb when day length is equal to or greater than 11-13 h. Intermediate cultivars bulb in response to day lengths equal to or greater than 13-14 h, and long-day cultivars bulb in response to day lengths of 14 h or longer. These designations have a positive correlation with latitude. For bulb production, short-day plants are usually grown at less than 30° latitude, intermediate be tween 30° and 38°, and those grown at latitudes greater than 38° are long-day types. All cultivars are long-day plants for the bulbing response, because they bulb in response to increasing rather than decreasing day length.

Photosynthate partitioning differs with various growth phases. During development prior to bulbing, leaf blade growth is greater than that of leaf sheaths. Induction of bulbing causes the mobilization of food reserves into the leaf bases, resulting in an enlargement that results in an increase in diameter at the base of the plant so that leaf sheath growth accelerates compared to leaf blade growth. Immediately after induction, successive new leaves tend to be longer and have wider leaf bases. Leaves and roots continue to be produced at a relatively uniform rate, although during bulbing, As bulbing advances, inner scale or bladeless leaf growth becomes dominant.

Bolting (Flowering)

Bolting is the formation of a seed stalk and associated inflorescence. Cultivars differ greatly in their response to low temperature and the duration of exposure needed for vernalization, the induction of bolting. A period of exposure to 5°-10°C for 4 to 6 weeks is adequate for the vernalization of many cultivars. Chilling degree days can be calculated from weather or storage room growth data using 10°C degrees as a base temperature for chilling hour accumulation. For some cultivars, temperatures between lo°C and 13°C are adequate to stimulate bolting. Insufficient vernalization results in low seed yields. It is possible to devernalize a plant by exposure to high temperatures subsequent to vernalization. The pattern of foliage growth and development is also altered after bolting (seed stalk development) is initiated because the developing seed stalk is a strong sink for photoassimilates. However. rapid vigorous bulbing can suppress seed stalk emergence even if it is already initiated as both processes compete for available photoassimilates. It is possible to have bulbs and seed stalks developing simultaneously. Sufficient plant size is required for flower induction to occur. Once beyond the juvenile seedling low temperatures can induce bolting. Large plants are more responsive to vernalization conditions than small ones. With less than four or five leaves or "neck" diameters less than 10 mm, less than a pencil diameter, are usually considered to be juvenile and not responsive. Plants grown for seed production should achieve ample leaf area before vernalization to support bolting. To get the vegetative growth needed for high seed yields, establishment should occur in mid to late summer so that large plants are sufficiently developed before inductive temperatures occur in the fall and winter. Seed stalks normally elongate as temperatures increase in spring after vernalization. In bulb-to-seed production, bulbs require a cool dormant period for floral primordia are initiation. The storage temperatures of bulbs influence sensitivity to bolting. Storage at either 0°C or 25°C is less conducive to bolting than temperatures in between. Subsequent seed stalk development is enhanced by large bulbs. The number of flower stalks per plant depends on the number of lateral bud shoot apices and large bulbs have more. However, when plants are grown directly from seed, usually only one seed stock is formed.

Roguing

Seed-to-seed

During autumn of first season, remove weeds, plants with off-type foliage, off-type bulb or stem color and plants bolting prematurely in first year. During flowering the following spring, check inflorescences and flower characters for gross abnormalities remove any weeds that have emerged since the previous roguing, rogue plants with off-type foliage or off-type bulb color .

Bulb-to-seed

Check inflorescences and flower characters for gross abnormalities. Flowering heads of plants infected with *Ditylenchus dipsaci* tend to be bent over; infected plants should be removed and burned. Before bulb maturity the field should be rogued to remove, plants with off-type foliage, off-type bulb or stem color, plants bolting prematurely in first year and late maturing plants. Early bolters, bull necks, bottle-shaped bulbs, split bulbs, doubles, damaged and diseased bulbs should be discarded prior to storage.

Pest Management

Managing pests in an important part of onion seed production. Both chemical and nonchemical strategies can be effective employed. Crop rotations to non-alliums for multiple years are a simple way to control many soil borne diseases, insect pests, certain weeds, and nematodes. Site history should be evaluated to make sure fields do not harbor noxious weed, insect or disease pests of onion. Areas near seed fields must also be managed to prevent alternate hosts from harboring pests which may migrate onto a seed crop. Weed control in areas adjacent production fields will reduce the introduction of weed seeds. Solarization and sustainable fumigation techniques should be considered if less invasive technique do not effectively control pernicious soil-borne pests.

Weed Management

Onions compete poorly with weeds because of the long period required for plants to achieve foliage cover. The long growing season required for onion seed crops allows for successive flushes of weed growth to occur. Since onion seed crops are planted in late summer and remain in the ground through the winter and spring in mild climates, they may encounter different types of weeds such as perennials, winter annuals, summer annuals, grasses and broadleaves. The limited availability of pre- and post-emergence herbicides makes site selection, preplant weed management, via cultivation an essential components of onion seed production. For annual broadleaf weed control, the stale bed technique may work effectively. For this technique, the bed is prepared as if for planting and irrigation is applied to encourage weed growth. After the first flush of weeds emerges, the weeds may be killed by shallow application of herbicides before the onion seed crop is planted. During winter months, cultivation may not be possible because of wet soils. Care must be taken when hand weeding because onion root systems may be damaged when weeds are removed close to plants. Two cycles of careful hand weeding along with careful site selection and management are necessary to keep weed populations low. Post emergence herbicides, where available, may be applied and the three to four leaf stage and later to provide effective control. Soil solarization prior to establishment is another weed control tool for areas with extended periods of intensity solar radiation. Morning glories and specifically field bindweed are particularly bothersome weeds because the seed is similar in size and shape and color to onion seed making it difficult to remove during post-harvest processing. In some countries, bindweed is a federally designated noxious weed meaning there is zero tolerance for contamination of onion seed lots.

Diseases

Disease problems encountered when producing onion seed are similar to those encountered for edible bulb production. Downy mildew is caused by the oomycete organism *Peronospora destructor*, which infects first the leaves and later bulbs of onions and other Alliums in mild, humid weather in spring and early summer as seed stalks develop. Downy mildew is often worse on direct-seeded crops. White rot (*Stromatinia cepivora* or syn. *Sclerotium cepivorum*) results from a soil-borne fungus, which can persist in the soil for many years and may be controlled by crop rotation. Purple blotch (*Stemphylium vesicarium* and *Alternaria porri*) are potentially serious fungal diseases that cause purple discoloration of foliage. Pink root (*Phoma terrestris*) and soil-borne fungus discolor roots and stunt onion plants. Neck rot (*Botrytis aclada* and *B. allii*) is primarily a postharvest storage disease that results in sunken and scales that turn gray to dark brown during storage for bulb-to-seed production. Proper curing can prevent this disease. Moist conditions favor botrytis leaf blight (*B. squamosa*). The disease causes leaf spots (lesions) and maceration of leaf tissue resulting in leaf dieback and blighting reducing onion growth and seed yield. Onion smut (*Urocystis cepulae*), a soil-borne seedling disease common in Northern areas, typified by distinctive narrow elongated dark streaks, usually on the cotyledon or first true leaf. Infected seedlings fail to emerge or usually die within a few weeks after emergence.

Insect Pests

Field scouting is important to identify troublesome insects as they appear. Both onion and flower thrips (*Frankliniella occidentalis*) feed on leaf sap, cause leaf distortion, reduce seed set, and by vectoring viruses such as IYSV. Thrips cause economic losses when they become too numerous on umbels. Onion maggots (*Delia antiqua*) feed on onion roots and overwinter in soil. Crop rotation can effectively control maggots. Corn seed maggot, wireworms, leaf miners, armyworms, mites, and cutworms may also attack onion seed fields. Broad spectrum chemical insecticides to control onion insect pests mustr be used with extreme caution, and not during bloom to protect insect pollinators at work in onion seed fields. Nematodes can also be a problem for onion seed production in some areas. The stem and bulb nematode (*Ditylenchus dipsaci*) and root knot nematodes (Meoidogye spp.) can stunt growth and reduce seed yields. Fields with a history of nematodes should be avoided.

Seed Harvest

Determining when to harvest is one of the most challenging aspects of onion seed production. This is largely because there are conflicting objectives: 1. To allow maximum seed maturation on the mother plant and 2. Minimize the loss of seeds from mature umbels due to shattering. The optimum time for seed harvest is based on cultivar characteristics and local weather. The seeds are black when ripe and can be clearly seen against the silvery colored capsules. Traditionally onion seed heads are harvested when about 10% of black seed are visibly exposed in the umbel. This visual stage of development corresponds to a seed or whole umbel moisture content of approximately 65%. Shattering increases sharply below an umbel seed moisture content of 50 to 55%. The seed heads shatter (seeds separate and fall from the head) if harvest is delayed so harvest should commence before 5% of seeds have been lost from the umbel. The optimum time for seed harvest is based on cultivar characteristics and local weather.

Harvesting may be direct or indirect. Direct harvesting occurs as a single operation in a once-over destructive harvest with a combine when the majority of field is ready for harvest as explained above. Significant seed losses occur in fields that lack uniformity but the advantage of direct harvest is that labor is greatly reduced. For direct harvesting, the beaters on the combine are removed to reduce shattering. Research shows that direct harvesting of the onion seed crop results in greater seed losses. The best time for mechanical harvesting with the least shattering losses is when the seeds are immature with a dry matter content of 60-70%. Adhesive materials have been applied to umbels experimentally to delay mechanical harvest until optimum seed quality is achieved and reduce shattering. However, such materials may foul harvest and conditioning equipment and add to production costs. In direct onion seed harvest remains popular with many companies despite its high labor costs because research shows that drying seeds while still in capsules attached to the stalks results in higher quality seed compared to seed dried in their capsules after separation from the umbels.

Indirect harvesting occurs when crops are hand harvested and dried prior to thrashing. For indirect harvesting, seed heads with approximately 10-20 cm of scape (seed stalk) attached are removed by cutting with a sharp knife or hand clippers. When cutting, the umbel is supported in the palm of the hand and held between the fingers to avoid loss of seed. Depending on crop uniformity and location, a single once-over hand harvest may be possible, but several successive hand harvests may be necessary for maximum seed yield. For indirect harvest, the seed heads are further dried on tarpaulins or sheets either in the open or in structures. For drying, umbels are placed 15 to 25 cm deep on a large canvas or plastic tarp and tried under ambient conditions for 2 to 3 weeks in areas without rainfall and low humidity. The piles are turned by hand to facilitate drying. Canvas tarps breath and are preferred to encourage air movement. Onion seed generally has a relatively short storage life and viability decreases rapidly at high temperatures and or higher seed moisture contents. To retain the highest seed quality, the seeds should dry quickly after harvest without an excessive buildup of heat on tarps. Frequent turning and shading will help maintain seed quality. Air circulation is key and fans can be employed to speed drying by lowering humidity. An alternative system designed to reduce labor cuts the stems at approximately 15 cm above ground level with a cycle-bar mower. The cut umbels with attached seed stocks are collected mechanically using a conveyer and placed in mobile bins. The material is then further dried on sheets as above. The attached stems add biomass to the piles increasing heating and requiring more frequent turning to facilitate drying prior to thrashing. In less arid regions, indoor drying may be necessary. Some producers use tiered boxes or a crate system to encourage air circulation. In climates with high humidity, forced-air drying or drying beads may be necessary. After drying, umbels or umbels with stalks are thrashed using conventional combines in the indirect system and partially cleaned (scalped) before transport to a milling facility for further processing.

Cleaning

Dried umbels are ready for threshing when the seeds can be separated from their capsules by rubbing in the hand. In order to avoid damaging the brittle seeds, threshing should not be delayed beyond this stage. Several threshing methods have been adopted for onion depending on the scale of operation and include flailing, rolling, threshing machines and combines. Onion seeds are very easily damaged during processing and frequent checks should he made to ensure that the seed coats are not accidentally cracked during processing. Examination of samples with a hand lens can confirm damage. Effect processing should separate seeds from the dried flowers without including pieces of the dried flower since this debris is difficult to separate from seed during cleaning. The light debris ("tailings") must also be examined to ensure it is free of good seed.

After thrashing, the seed is transported to a cleaning and milling facility for further processing. When seed arrives at a processing plant the moisture content should be 8 or 9% to prevent heat buildup in bulk storage. If the moisture content is above this threshold, seeds should be dried using forced air. Raw seed is milled to remove debris, which may account for as much as 10 to 20% of the seed lot by weight. After threshing, the initial cleaning is often achieved using an air-screen machine. Density gravity tables can further upgrade onion seed lots. If the seed is not sufficiently clean after these operations, or if it contains noxious weed seed, the lot may be run through the previous steps again or more specialized equipment used.

Flower pedicels may be removed by either a magnetic separator (the iron powder is added which adheres to the pedicels and not the seeds) or by flotation. If the latter process is used, only the light fraction is placed in water to avoid seed damage. During this process, which should not exceed three minutes per seed-lot, the good seeds sink while the poor-quality light seeds and pedicels float off. After spin drying and further drying in racks, additional debris is removed by an air-screen cleaner. Seed cleaning removes some good seed as well as contamination, so only the minimum number of steps required to meet the desired purity standards are used. The final seed lot must dried to 8% moisture content or less depending on the method of storage and packaging.

After cleaning samples are analyzed for moisture, germination, and purity. If results are substandard, further milling may be required to upgrade the lot. Once minimum germination is achieved (commonly 85% in many countries), moisture content is again checked and adjusted below 8% before the seed is stored for packaging. Seeds may be treated with various materials such as biologicals, minerals, coating materials etc. and packaged in metal cans or plastic buckets for larger quantities or foil or plastic packets for smaller quantities. Since onion has a relative short shelf life, seeds must be protected from heat and humidity during storage and shipment using temperature controlled storage and resealable packaging with a barrier to keep seed moisture low.

Seed Yield

The best yield from open-pollinated crops produced under ideal conditions is 2000 kg/ha but 1000 kg/ha is more common. In some areas yields as low as 500 kg/ha may be deemed satisfactory. The yield from F-1 hybrids is normally lower than from open-pollinated crops and is often as low as 50-100 kg/ha. There are approximately 9000 onion and leek seeds in one ounce.

Shallot

Shallots, *Allium cepa*, are closely related to multiplier onions, but smaller, and have unique culinary value. (The term 'multiplier' means that the bulbs multiply freely producing several lateral bulbs). At maturity, shallot bulbs resemble small onions.

Shallots have long been associated with fine French cuisine. They are eaten fresh or cooked, chopped or boiled. Shallots have a delicate onion flavour when cooked that adds to but does not overpower other flavours.

Figure: Shallot plant made up of a multitude of small bulbs

Raw shallots have a strong pungency, stronger than most onions. Their true character comes when lightly sauteed in butter until they are translucent in colour or when used in gravies and creamy sauces. It is very difficult to evaluate shallot quality in the raw form.

Shallots can be successfully produced wherever onions are grown. However, most shallots are produced in Europe, particularly France.

Most shallots consumed in the USA and Canada are imported chopped and dried from Europe. Otherwise, those that are used fresh are consumed green, much like that of green bunching onions, since the mature bulb of the shallot is small.

The difference between a multiplier onion and shallot is somewhat arbitrary, and they are often lumped together. Commercially however, those with yellow or brown scales and white interiors, such as the 'Dutch Yellow' type, are usually classed as multiplier onions, while those with red scales and, supposedly, a distinctive and more delicate flavour, are classed as shallots.

Taxonomically, there is no such thing as a true shallot.

Many people confuse shallots with green onions, scallions and leeks. The young green onion has a definite bulb formation with the same concentric arrangement that the dry onion has. Scallions are any shoots from the white onion varieties that are pulled before the bulb has formed. Leeks are similar in appearance to scallions but have flat leaves and the white stalk is thicker and longer. The shallot can be distinguished from the others by its distinctive bulbs which are made up of cloves like garlic, but unlike garlic, the individual bulbs are not encircled together by a common membrane.

Generally, shallot bulbs are the size of chestnuts, sometimes larger, pear-shaped, narrowed in the upper part into a rather longpoint, and covered with a russet coloured skin of a coppery red colour in the lower part shading off into grey towards the upper extremity.

In the grey shallot, which is sometimes claimed to be the "true shallot", the scales or skin are often wrinkled length-wise and are thick and tough. When the dried skin is taken off, the bulb is often greenish at the base and violet coloured toward the top. The roots are slow drying and persistent. Leaves are small, very green, and 4–5 cm long.

Propagation and Cultural Management

Shallots can be grown from seed, but usually small bulbs are planted in late fall or early spring. The "mother" bulbs divide forming several bulbs.

Although shallots are mostly thought of as dry bulbs, in some areas the green shoots of shallots are used similarly to the green onion or as a scallion substitute.

Plant the bulbs 10–15 cm (4–6") apart. The size of the bulb affects the date of sprouting, plant size and maturity. For uniformity in production, planting similar size bulbs is essential.

For early maturity and harvest, strong, healthy transplants can be used. Transplants can be started 30 to 45 days before direct bulb seeding in the field, and plants can be moved to the field in 30 to 60 days. Don't plant the bulbs or plants deeply and do not move soil to cover the plant base; the bulbs should grow out of the ground for easier dividing.

To harvest over an extended period, plant the largest bulbs (quarter-size) first. After they mature, plant medium-sized bulbs (nickel-size), and finally the smallest bulbs (dime-size). If using transplants, plant one large or two small plants in each hole. Discard the weak clumps and the smallest plants.

To save bulbs for the following year, save only the highest quality bulbs from the highest quality clumps. Many growers market the biggest bulbs and save smaller bulbs for replanting; but this results in gradually smaller and poorer quality shallots. Save bulbs from the biggest and best clumps. These clumps should be as free from disease as possible.

Figure: Shallot bulbs cured and ready for market

Shallots need a continuous supply of nutrients. Split or continuous applications of nitrogen are essential for good growth. Approximately, 75–100 kg/ha of total nitrogen is required for this crop, and its application much like for that of garlic.

The root system is weak and shallow, thus irrigation in the spring and summer is more frequent than for other vegetables. Soil type does not effect the total amount of water needed, but does dictate frequency of water application. Lighter soils need more frequent water applications, but less water applied per application.

Green shallots can be harvested in 30 to 60 days, mature bulbs in 90 to 120 days. Pre-harvest, harvest, and post-harvest management of shallots is similar to that of onions.

Quality and storageability is enhanced by low nitrogen late in the season, proper and adequate field curing, removing tops only after they are dried, and storing the bulbs in a cool, dry area.

One thousand kilograms of seed bulbs should yield 5,000 to 7,000 kilograms of shallots.

Storage

Shallots store well at temperatures of 0–2°C and 60%–70% relative humidity. Because of their small size, shallots tend to pack closely; so they should not be placed into deep piles. Store shallots in slatted crates or trays that allow good air movement in and around the bulbs. This is important to remove excessive moisture and to minimize storage diseases.

Low relative humidity and low temperatures are important to keep shallots sound and dormant and free from sprouting and root growth. At humidities much above 70% and at warmer temperatures of 5–8°C more of the shallots will sprout, develop roots, and decay. With good air flow and humidity control, shallots should store for 8–10 months.

Varieties

Two general types of shallots are available. French-Italian has brownish-red skin, well-shaped bulbs, and a flavour between onion and garlic. Varietal names include Pikant, Atlas, Ambition, Ed's Red, and Creation. They can be used either dry or green.

A second type is Welsh shallot. Louisiana Evergreen is one variety. It bulbs poorly, but provides a year-round supply of green shallots for salads, seasonings, or appetizers.

The red shallot is essentially the only one of importance in the market place.

Pests and Diseases

Shallots are susceptible to bacterial diseases, pink root, white rot, downy mildew, purple blotch, onion maggot and thrips. To avoid or minimize these problems, do not plant shallots in the same soil where other Alliums have been grown in recent years, plant only clean, healthy plants or bulbs, and practice good sanitation.

Garlic

Garlic is a bulbous plant belonging to Amaryllidaceae family. Shallots, onion, leek etc. belong to this family. Scientific name of garlic is Allium sativum. Garlic plant grows to a height of 4 ft. and produces flowers. It can be propagated both sexually as well as vegetatively. For cultivation purposes, garlic is propagated asexually by sowing the cloves. There are different varieties of garlic for different use.

Climate for Garlic Farming

Garlic cultivation needs a combination of different types of climate. It needs a cool and moist climate for bulb development and vegetative growth while for maturity the climate must be warm and dry. However, it cannot tolerate extreme cold or hot conditions. Exposing the young plants to temperatures lower than 20°C for 1 or 2 months would hasten the bulb formation. A prolonged exposure to lower temperature would however reduce the yield of the bulbs. Bulbs maybe produced at the axil of the leaves. A cooler growing period gives higher yield than warmer growth conditions. The optimal day length requirement for bulb formation is 13-14 hours for long day garlic and 10-12 hours for short day garlic.

Soil for Growing Garlic

Although garlic can grow in different types of soil, loamy soil with natural drainage is optimum for this crop. It grows at an altitude of 1200 to 2000 m above sea level. It is sensitive to acidic and alkaline soils, hence, a pH of 6-8 is suitable for optimal growth of garlic. A clayey, water-logging type soil is also not suitable for garlic growing. Soils with rich organic content, good moisture, high amount of nutrients aid in proper bulb formation. A heavy soil with less moisture and more water logging would result in deformed bulbs. Soils with poor drainage capacity causes discolored bulbs.

Irrigation

Garlic is a bulb crop producing shallow roots. It therefore, requires a good amount of moisture-more than water. Perhaps the biggest challenge in garlic cultivation is being able to 'moisture it right'. In other words, there should be enough water to maintain a good level of moisture in soil. However, too much water would result in water stress and thus splitting of the bulbs. Too little water or moisture level again means under-developed bulbs. The best way is to irrigate the crop frequently. It must be irrigated:

- Immediately after planting
- At an interval of one week to 10 days depending on the moisture content in soil.

Alternating the irrigation period with a dry spell causes the outer scales of garlic to split. Water-logging results in development of diseases like purple blotch and basal rot. A continuous irrigation until maturity causes secondary roots to develop. Such crops produce new sprouts and growth. Bulbs from these crops cannot be stored for a long period of time.

The best way to irrigate garlic is by use of modern day techniques like sprinkler and drip irrigation. It helps in improving the yield considerably. In case of drip irrigation, the discharge flow rate of the emitters must be 4liters per hour. It helps improve the yield by 15-25% better than flood irrigation system. The discharge rate in sprinklers must be 135 liter per hour.

Fertigation in Garlic Cultivation

Fertigation is an efficient method of using drip irrigation to apply fertilizers. The drip emitters are used as carriers of both water and crop nutrients. At the time of planting a basal dose of 30 Kg nitrogen per acre is applied. Applying nitrogenous fertilizers through drip irrigation is more efficient because the nutrients are directly applied to the root zone. Loss of nitrogen through ground water leaching is reduced.

Crop Rotation

Garlic is a shallow-rooted crop. Therefore, it would not utilize all the nutrients supplied. These fertilizers and nutrients leach down with water and settle in the sub-soil. It can be used by deep-rooted leguminous crops. Research has shown that alternating garlic with leguminous crops not only improve the boosts the yield of garlic but also improves the soil fertility. Alternating crops like groundnut with garlic can ensure better profit for the farmers as well.

Land Preparation

Loose and well-drained soil is a must in garlic cultivation since moisture is one of the pre-requisites of garlic plant. Therefore, the land to be used for plantation must be well-ploughed, free of clods and other debris. It is advisable to use a moldboard plough since it enhances the drainage property of the soil and also pushes the crop residue deeper down the earth to enhance better decomposition. In order to get rid of the soil clod, the land is tilled 3-4 times with organic manure being incorporated at the last time. For rabi crops, flat beds of 4-6 m length and 1.5-2 m width are formed. However, flat beds are avoided during the kharif or rainy seasons so as to prevent water logging. In case of kharif crops, broad bed furrows (BBF) with a height of 15 cm are made. The top width is about 120 cm and each furrow is 45 cm deep. Broad Bed Furrows are suitable for drip and sprinkler irrigation. The rows must be made at a distance of 15 cm from each other.

Methods to Plant Garlic

Individual cloves used for planting garlic must be separated. However, the basal plate of the clove must be undamaged as that is the place from where roots develop. Typically, bigger cloves are used for garlic planting while smaller cloves are rejected. Some people use the smaller, rejected cloves for pickling. Cloves to be used for planting must be dipped in 0.1% carbendazim solution just before sowing. This reduces the incidence of fungal diseases. They are then planted perpendicular to the ground. The distance between two garlic plants must be at least 10 cm.

Diseases and Plant Protection

Looking after a crop and protecting it from diseases is the biggest task in a farmer's life. Disease and pest management is vital for obtaining good quality yield of bulbs. There are different types of diseases in garlic caused by viruses, fungi, nematodes and insects.

Viral Diseases

Name of Disease	Causative Agent	Symptoms and Nature of Damage	Spread By	Control
Onion Yellow Dwarf disease	Onion Yellow Dwarf Virus	• Leaves develop yellow streaks at the base of true leaves and subsequent leaves • They crinkle and fall off. • Bulbs are under-sized	Aphids, infected cloves and seed bulbs can transmit the disease.	• Using healthy seed material • Foliar spraying of carbosulfan (0.2%) or fipronil (0.1%) to control aphids is one of the measures.
Leek Yellow Stripe	Leek Yellow Stripe Virus	• Yellow stripes on the distal part of the leaves • Produces malformed bulbs	Aphids	• Using healthy seed material • Foliar spraying of carbosulfan (0.2%) or fipronil (0.1%) to control aphids is one of the measures.
Irish Yellow Spot	Irish Yellow Spot Virus	• Spindle-shaped, straw-colored spots on the leaves • Poorly defined edges of the leaves • Spots coalesce becoming bigger	Onion thrips and transplantation of infected plant	• Avoid crop stress • Control thrips • Use healthy planting material

Fungal Diseases

Disease	Causative Agent	Symptom and Nature of Damage	Transmission	Control
Purple Blotch	Alternaria porri	• Elliptical, small lesions on leaves • Lesions turn purplish-brown and are surrounded by a chlorotic margin • Lesions start at the tip of older leaves • They girdle the leaves and hence leaves fall off	It is chiefly soilborne disease also spread by infected bulbs, plant debris, etc.	Spraying 0.25% mancozeb, 0.1% propiconazole or 0.1% hexaconazole every 15 days after 30 days of planting or immediately when symptoms appear would help control the disease.
Stemphylium blight	*Stemphylium vesicarium*	• Yellow to orange colored streaks develop on the leaf • The streaks develop into elongated or spindle-shaped spots characterized by pink-colored margins. • Spots coalesce from tip to base of the leaves	Plant debris and soil are the major transmitting agents	Spraying 0.25% mancozeb, 0.1% propiconazole or 0.1% hexaconazole every 15 days after 30 days of planting or immediately when symptoms appear would help control the disease.

White rot	*Sclerotium cepivorum*	• Yellowing and dying of leaf tips • Sclerotia on the surface of the leaves or within the tissues	Soil, garlic debris and diseased garlic sets	• High temperature soil solarization • Destroying the infected crop • Crop rotation • 0.1% carbendazim should be used

Insect Pests

Name of Pest	Identification	Symptom	Control
Thrips tabaci	• Body color varies from yellow to dark brown • 4 wings with long hairs	• Leaves twist and curl in the initial stages • Silvery to white patches on leaves • Plants seem blemished and turn white in extreme cases	• Planting one or two outer rows of maize or wheat can act as a good barrier for thrips. This is typically done 30 days before planting garlic. • Spraying carbosulfan (0.2%), fipronil (0.1%)or profenofos (0.1%) when thrips population cross 30 thrips per plant is advisable.
Eriophyid mite (*Aceria tulipae*)	Identification with naked is difficult as they are microscopic, banana-shaped organisms.	• Leaves do not open fully • Plant shows curling, twisting and stunting • Mottling along the edges of leaves	• Spraying Sulphur powder as soon as the symptoms appear and every fortnight thereafter would help check the progress of the disease.

Harvesting Garlic

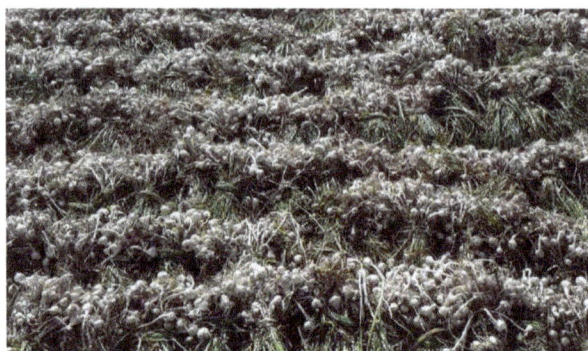

Harvesting from garlic cultivation

Garlic is ready to harvest within 120-150 days of sowing depending on the variety. They are ready when the leaves start yellowing and become dry. The bulbs are then pulled out, sheath cut near the bulb and roots are trimmed. They are then sun-dried for a week. This process is important for the hardening of the bulbs. Before storing they are graded according to the size and weight.

Garlic Seed Storage

Garlic can be stored at room temperature of up to 8 months. Before storing it must be sun dried thoroughly so that no fungus develops on it during the storage period.

Storing garlic

Modified Plant Stem Vegetables

Potato

Potato is one of the most important commercial crops that is cultivated widely because of its great demand in the market, throughout the whole year. Because of plentiful benefits of potatoes, there is great demand for potato products in both local and international market.

It is essentially a "cool weather crop", with temperature being the main limiting factor on production: tuber growth is sharply inhibited in temperatures below 10°C (50° F) and above 30°C (86° F), while optimum yields are obtained where mean daily temperatures are in the 18 to 20°C (64 to 68° F) range.

For that reason, potato is planted in early spring in temperate zones and late winter in warmer regions, and grown during the coolest months of the year in hot tropical climates. In some sub-tropical highlands, mild temperatures and high solar radiation allow farmers to grow potatoes throughout the year, and harvest tubers within 90 days of planting (in temperate climates, such in northern Europe, it can take up to 150 days).

The potato is a very accommodating and adaptable plant, and will produce well without ideal soil and growing conditions. However, it is also subject to number of pests and diseases. To prevent the build-up of pathogens in the soil, farmers avoid growing potato on the same land from year to year. Instead, they grow potato in rotations of three or more years, alternating with other, dissimilar crops, such as maize, beans and alfalfa. Crops susceptible to the same pathogens as potato (e.g. tomato) are avoided in order to break potato pests' development cycle.

With good agricultural practices, including irrigation when necessary, a hectare of potato in the temperate climates of northern Europe and North America can yield more than 40 tonnes of fresh tubers within four months of planting. In most developing countries, however, average yields are much lower - ranging from as little as five tonnes to 25 tonnes - owing to lack of high quality seed and improved cultivars, lower rates of fertilizer use and irrigation, and pest and disease problems.

Soil and Land Preparation

The potato can be grown almost on any type of soil, except saline and alkaline soils. Naturally loose soils, which offer the least resistance to enlargement of the tubers, are preferred, and loamy and sandy loam soils that are rich in organic matter, with good drainage and aeration, are the most suitable. Soil with a pH range of 5.2-6.4 is considered ideal.

Growing potatoes involves extensive ground preparation. The soil needs to be harrowed until completely free of weed roots. In most cases, three ploughings, along with frequent harrowing and rolling, are needed before the soil reaches a suitable condition: soft, well-drained and well-aerated.

Planting

The potato crop is usually grown not from seed but from "seed potatoes" - small tubers or pieces of tuber sown to a depth of 5 to 10 cm. Purity of the cultivars and healthy seed tubers are essential for a successful crop. Tuber seed should be disease-free, well-sprouted and from 30 to 40 g each in weight. Use of good quality commercial seed can increase yields by 30 to 50 percent, compared to farmers' own seed, but expected profits must offset the higher cost.

The planting density of a row of potatoes depends on the size of the tubers chosen, while the inter-row spacing must allow for ridging of the crop. Usually, about two tonnes of seed potatoes are sown per hectare. For rainfed production in dry areas, planting on flat soil gives higher yields, while irrigated crops are mainly grown on ridges.

Stages in crop development

1. Planted seed tuber
2. Vegetative growth
3. Tuber initiation
4. Tuber bulking

Crop Care

During the development of the potato canopy, which takes about four weeks, weeds must be controlled in order to give the crop a "competitive advantage". If the weeds are large, they must be removed before ridging operations begin. Ridging (or "earthing up") consists of mounding the soil from between the rows around the main stem of the potato plant. Ridging keeps the plants upright and the soil loose, prevents insect pests such a tuber moth from reaching the tubers; and helps prevent the growth of weeds.

After earthing up, weeds between the growing plants and at the top of the ridge are removed mechanically or using herbicides. Ridging should be done two or three times at an interval of 15 to 20 days. The first should be done when the plants are about 15-25 cm high; the second is often done to cover the growing tubers.

Manuring and Fertilization

The use of chemical fertilizer depends on the level of available soil nutrients - volcanic soils, for example, are typically deficient in phosphorus - and in irrigated commercial production, fertilizer requirements are relatively high. However, potato can benefit from application of organic manure at the start of a new rotation - it provides a good nutrient balance and maintains the structure to the soil. Crop fertilization requirements need to be correctly estimated according to the expected yield, the potential of the variety and the intended use of the harvested crop.

Water Supply

The soil moisture content must be maintained at a relatively high level. For best yields, a 120 to 150 day crop requires from 500 to 700 mm (20 to 27.5 inches) of water. In general, water deficits in the middle to late part of the growing period tend to reduce yield more than those in the early part. Where supply is limited, water is directed towards maximizing yield per hectare rather than being applied over a larger area.

Because the potato has a shallow root system, yield response to frequent irrigation is considerable, and very high yields are obtained with mechanized sprinkler systems that replenish evapotranspiration losses every one or two days. Under irrigation in temperate and subtropical climates, a crop of about 120 days can produce yields of 25 to 35 tonnes/ha (11 to 15.6 tons per acre), falling to 15 to 25 tonnes/ha (6.6 to 15.6 tons per acre) in tropical areas.

Pests and Diseases

Against diseases, a few basic precautions – crop rotation, using tolerant varieties and healthy, certified seed tubers - can help avoid great losses. There is no chemical control for bacterial and viral diseases but they can be controlled by regular monitoring (and when necessary, spraying) of their aphid vectors. The severity of fungal diseases such as late blight depends, after the first infection, mainly on the weather - persistence of favourable conditions, without chemical spraying, can quickly spread the disease.

Insect pests can wreak havoc in a potato patch. Recommended control measures include regular monitoring and steps to protect the pests' natural enemies. Even damage caused by the Colorado Potato Beetle, a major pest, can be reduced by destroying beetles, eggs and larvae that appear early in the season, while sanitation, crop rotations and use of resistant potato varieties help prevent the spread of nematodes.

Harvesting

Yellowing of the potato plant's leaves and easy separation of the tubers from their stolons indicate that the crop has reached maturity. If the potatoes are to be stored rather than consumed immediately, they are left in the soil to allow their skins to thicken - thick skins prevent storage diseases and shrinkage due to water loss. However, leaving tubers for too long in the ground increases their exposure to a fungal incrustation called black scurf.

To facilitate harvesting, the potato vines should be removed two weeks before the potatoes are dug up. Depending on the scale of production, potatoes are harvested using a spading fork, a plough or commercial potato harvesters that unearth the plant and shake or blow the soil from the tubers. During harvesting, it is important to avoid bruising or other injury, which provide entry points for storage diseases.

Storage

Since the newly harvested tubers are living tissue – and therefore subject to deterioration - proper storage is essential, both to prevent post-harvest losses of potatoes destined for fresh consumption or processing, and to guarantee an adequate supply of seed tubers for the next cropping season.

For ware and processing potatoes, storage aims at preventing "greening" (the build up of chlorophyll beneath the peel, which is associated with solanine, a potentially toxic alkaloid) and losses in weight and quality. The tubers should be kept at a temperature of 6 to 8°C degrees, in a dark, well-ventilated environment with high relative humidity (85 to 90 percent). Seed tubers are stored, instead, under diffused light in order to maintain their germination capacity and encourage development of vigorous sprouts. In regions, such as northern Europe, with only one cropping season and where storage of tubers from one season to the next is difficult without the use of costly refrigeration, off-season planting may offer a solution.

Yam (Vegetable)

Amorphophallus paeoniifolius, the elephant foot yam or whitespot giant arum or stink lily, is a tropical tuber crop grown primarily in Africa, South Asia, Southeast Asia and the tropical Pacific islands. Because of its production potential and popularity as a vegetable in various cuisines, it can be raised as a cash crop. Elephant foot yam is of Southeast Asian origin. It grows in its wild form in Sri Lanka, the Philippines, Malaysia, Indonesia, and other Southeast Asian countries.

Soil

A rich red-loamy soil with a pH range of 5.5-7.0 is preferred. It is a tropical and subtropical crop. It requires well distributed rainfall with humid and warm weather during vegetative phase and cool and dry weather during the corm development period.

Season and Planting

It undergoes a dormancy period of 45 to 60 days. Traditionally farmers take advantage of the dormancy period by planting during February-March so that the setts would sprout with the pre-monsoon showers. April – May is the planting season. The tuber is cut into 750-1000 g small bits in such a way that each bit has atleast a small portion of the ring around each bud. Whole corms of

500 g size can also be used as a planting material. Use of cormels and minisett transplants of 100 g size as planting material at a closer spacing of 45 x 30 cm is also suggested. There are also projections with tender buds called "Arumbu". These are removed before planting as they do not give vigorous growth. An ordinary sized yam gives about 6 to 8 bits for planting. The cut pieces are dipped in cow dung solution to prevent evaporation of moisture from cut surface. In some places, the small round daughter corms are also planted. The cut pieces are planted in beds at 45 cm x 90 cm spacing or pit of 60 x 60 x 45 cm size is dug and planted. The pit should be filled with top soil and farm yard manure (2 kg/pit) prior to planting. The pieces are planted in such a way that the sprouting region (the ring) is kept above the soil. About 3500 kg of corms will be required to plant one hectare. Sprouting takes place in about a month.

Preparation of Field

The land is brought to fine tilth and form beds of convenient size.

Planting

The cut pieces are planted in beds at 45 cm x 90 cm spacing. The pieces are planted in such a way that the sprouting region (the ring) is kept above the soil. Sprouting takes place in about a month.

Intercropping

Vegetable cowpea var. CO_2 is recommended as suitable intercrop in elephant foot yam. It can be intercropped profitably in coconut, arecanut, rubber, banana and robusta coffee plantations at a spacing of 90 x 90 cm. Half quantity of FYM (12.5 t/ha) and one third of NPK (27:20:33) will be sufficient for the intercrop.

Irrigation

It is mostly raised as a rainfed crop. However, irrigation is required when monsoon fails, where it is grown on a large scale. Water stagnation is harmful to the crop. Wherever irrigation facility is available, irrigation can be given once a week.

Application of Fertilizers

Apply 25 tonnes of FYM/ha during last ploughing. The recommended dose of NPK/ha is 80:60:100 kg. Apply 40:60:50 kg NPK/ha at 45 days after planting along with weeding and intercultural operations. Top dress with 40:50 N and K one month later along with shallow intercultural operations.

After Cultivation

Weeding and earthing up as and when necessary.

Plant Protection

Disease

Leaf Spot

Leaf spot disease can be controlled by spraying Mancozeb at 2 g/lit

Leaf spot disease

Collar Rot

The disease is caused by a soil borne fungus Schlerotium rolfsii. Water logging, poor drainage and mechanical injury at collar region favour the disease incidence. Brownish lesions first occur on collar regions, which spreads to the entire pseudostem and cause complete yellowing of the plant. In severe case, the plant collapses leading to complete crop loss.

Management

Use disease free planting material, remove infected plant materials, improve drainage conditions, incorporate organic amendments like neem cake, drench the soil with carbenilazim or apply bio-control agents like Trichoderma harzianumI 2.5 kg/ha mixed with 50kg of FYM (lg/l of water).

Harvest

Harvesting is done on 8 months after planting and particularly during January - February months. Drying of stem and leaves indicates the harvesting stage in elephant yam.

Maranta Arundinacea

Arrowroot (*Maranta arundinacea* L.) is a tropical herb used for its tubers, which contain a highly valuable starch. The leaves and the by-products of starch extraction are fed to livestock.

Morphology

Arrowroot is a perennial glabrous herb that can grow up to 1.5 m in height. It forms thickets in shaded places. It has shallow roots and many cylindrical fleshy tubers (rhizomes) that can go deeper in the soil than the roots. Arrowroot is many branched. Its stems are slender and each bears one large (10-25 cm long x 3-8 cm broad) ovate and oblong leaf at its end. The flowers are small, white in colour, and borne in panicles.

Utilisation

Arrowroot is mainly cultivated for its starchy tubers. Tubers must be processed within 48 hours of harvest because they are prone to rotting. After being soaked in hot water, they are peeled to remove their fibrous covering in order to prevent a bitter taste and discolouration in the final product. They are then cut into small pieces and grated into a coarse pulp. Macerating the pulp breaks down the tough cells surrounding the starch. The pulp is then washed on screens to separate the starch from the fibrous material. The settled starch is then centrifuged or filtered to further separate it from fibre fines and other soluble material. The separated starch must be quickly dried and ground to powder. The root contains about 20% starch, and 17-18% can be extracted. Arrowroot starch is one of the purest natural carbohydrates, with a high maximum viscosity. It is used in food preparations and confectionery, such as jellies and pastries. Since it is highly digestible, it is fed to infants and to people with specific dietary requirements. Industrial applications include cosmetics and glue. It has several ethnomedicinal applications.

Production of starch from arrowroot yields several potential feed resources. The aerial biomass of arrowroot that is left in the field can be processed into silage. The residue of starch extraction, called "bittie" in the West Indies island of Saint Vincent has a high fibre content and can be fed to livestock.

Forage Management

Arrowroot starts flowering about 90 days after planting, and the rhizomes are mature after 10-11 months. Harvesting operations can start when foliage starts to wilt. Successive crops are usually grown on the same land for 5-6 years. Root harvesting is usually done manually. The plants are dug up and the tubers separated from the leafy stems. Mechanical harvesting was reported to be difficult. Yields range from 10 to 35 t/ha of rhizomes, from which 2.5-7.5 t/ha of starch can be obtained.

True Root Vegetables

Taproot

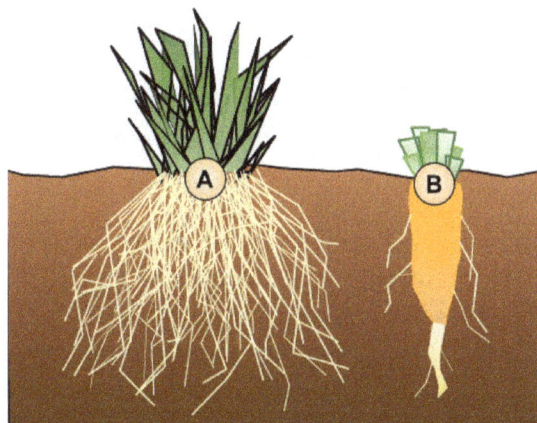

The two types of root systems in plants. The fibrous root system (A) is characterized by many roots with similar sizes. In contrast, plants that use the taproot system (B) grow a main root with smaller roots branching off of the taproot. The letters mark the beginning of the roots

A taproot is a large, central, and dominant root from which other roots sprout laterally. Typically a taproot is somewhat straight and very thick, is tapering in shape, and grows directly downward. In some plants, such as the carrot, the taproot is a storage organ so well developed that it has been cultivated as a vegetable.

The taproot system contrasts with the adventitious or fibrous root system of plants with many branched roots, but many plants that grow a taproot during germination go on to develop branching root structures, although some that rely on the main root for storage may retain the dominant taproot for centuries, for example *Welwitschia'*.

Development

Figure: A dandelion taproot (left) with the plant (right)

Figure: The edible, orange part of the carrot is its taproot

Taproots develop from the radicle of a seed, forming the primary root. It branches off to secondary roots, which in turn branch to form tertiary roots. These may further branch to form rootlets. For most plants species the radicle dies some time after seed germination, causing the development of a fibrous root system, which lacks a main downward-growing root. Most trees begin life with a taproot, but after one to a few years the main root system changes to a wide-spreading fibrous root system with mainly horizontal-growing surface roots and only a few vertical, deep-anchoring roots. A typical mature tree 30–50 m tall has a root system that extends horizontally in all directions as far as the tree is tall or more, but as much as 100% of the roots are in the top 50 cm of soil.

Soil characteristics strongly influence the architecture of taproots; for example, deep rich soils favour the development of vertical taproots in many oak species such as *Quercus kelloggii*, while clay soils promote the growth of multiple taproots.

Horticultural Considerations

Many plants with taproots are difficult to transplant, or even to grow in containers, because the root tends to grow deep rapidly and in many species comparatively slight obstacles or damage to the taproot will stunt or kill the plant. Among weeds with taproots dandelions are typical; being deep-rooted, they are hard to uproot and if the taproot breaks off near the top, the part that stays in the ground often resprouts such that, for effective control, the taproot needs to be severed at least several centimetres below ground level.

Sweet Potato

Sweet potato is cultivated as a perennial in tropical and subtropical lowland agro- ecologies, although it is well adapted to other zones and can be grown over widely different environments.

Climatic Requirements

Temperature

Because sweet potatoes are of tropical origin, they adapt well to warm climates and grow best during summer. Sweet potatoes are cold sensitive and should not be planted until all danger of frost is past. The optimum temperature to achieve the best growth of sweet potatoes is between 21 and 29°C, although they can tolerate temperatures as low as 18°C and as high as 35°C. Storage roots are sensitive to changes in soil temperature, depending on the stage of root development.

Soil Requirements

Site Selection and Soil

A well-drained sandy loam is preferred and heavy clay soils should be avoided as they can retard root development, resulting in growth cracks and poor root shape. Lighter soils are more easily washed from the roots at harvest time. Wet season green manure cropping with sterile forage sorghum is recommended and should be thoroughly incorporated and decomposed by planting time.

Soil pH should be adjusted to about 6,0 by applying lime or dolomite. Rates of 240 kg and 400 kg/ha respectively will raise the pH by 0,1 of a unit. The soil should be deep ripped and then disk cultivated to break up any large clods and provide enough loose soil for hilling of beds. A yearly soil test is recommended to assess soil properties, pH and nutrient levels before ground preparation.

Cultivation Practices

Propagation

Sweet potatoes are propagated from sprouts or from slips (vine cuttings); sprouts are preferred.

Sprouts are grown from plant stock selected for its appearance, freedom from disease and off-types. Approximately 75 kg of planting stock sweet potatoes are needed to produce enough sprouts to plant one hectare.

Cutting Collection

Tip cuttings of about 30 to 40 cm long with approximately eight nodes are collected from the nursery bed, or the last established planting. Tip cuttings should be taken from crops that are old enough to provide material without excessive damage. Avoid "back cuts" as these will have variable maturity and result in significant yield reduction. The lower leaves should be cut away as tearing these off may damage the nodes that will produce the roots. Cuttings can be left under a moist cloth in the shade for a couple of days to promote nodal rooting before planting in the field. At the recommended plant spacing, 330 cuttings are required for a 100 m row.

Seedbed Production of Cuttings

This involves the propagation of cuttings from harvested roots which are placed close together in a seedbed. This is an alternative method of producing planting material which requires less labour but does sacrifice a percentage of marketable roots. Research was conducted at the Coastal Plains Horticulture Research Farm (CPHRF) in 1992. Seedbed production of cuttings showed that it required about 25 kg of roots to plant 1 m² of seedbed, which yielded approximately 200 cuttings per cut over four cuts.

Planting Cuttings

Cuttings should be planted at about a 450 angle into heaps as this promotes good, even root development. Half of the cutting or three to four nodes should be buried at a spacing of 30 cm between plants. Mechanical planters are available and used on large-scale plantings but manual planting is widely practised. This can be as easy as pushing the cutting into the heap with a forked stick. The labour requirement for hand planting is estimated at 32 h/ ha. Cuttings need to be watered at or immediately after planting. Plantings should be scheduled to allow for progressive fortnightly harvests over the desired production period.

Sprout Production

Sprouts are produced from the conditioned roots in cold frames, heated beds, or field beds of clean sand or fumigated sandy soil. Conditioned roots are covered by more soil sand, though not too much. Four to five weeks are needed to develop strong plants if the soil in the plant beds has been kept at 23 to 26°C. Six to eight weeks may be needed if roots have not been "preconditioned". Adequate moisture is especially critical to germination of the sprouts and proper root formation on the sprouts.

Planting the Sprouts

Sprouts should be taken from the plant beds when 6 to10 leaves and a strong root system have developed on each one. They are set out into the field as early as possible when the soil has warmed and the risk of frost or a cold weather period has passed.

Plants should be spaced 30 to 38 cm apart in rows that are 1 m apart. This requires approximately 14 520 plants per hectare. Management of water is very critical to avoid transplant shock.

Soil Preparation

Bed Formation

Sweet potato is grown on raised beds or mounds. This provides the developing roots with loose, friable soil to expand to their potential size and shape without restriction. It also allows adequate drainage and provides easy harvesting with a mechanical digger.

Mounds should be approximately 30 cm high and 40 cm wide at the base. The main consideration is that the developing roots remain under the soil within the heaps. If using a mechanical digger at harvest time it is important to match the width of the mound with the width of the digger mouth. Spacing the mounds at 1.5 to 2.0 m apart (depending on the tractor width) with a roadway every six rows allows access for boom spray. Mounds are formed, using hilling discs, and the base fertiliser can be incorporated during this operation.

Planting

Planting Period

Planting time is mainly determined by the climate of a location. Sweet potato plants are damaged by light frost and the plants require high temperatures for a period of 4 to 5 months to yield well.

In areas with mild frost, mid-November to mid-December is the best time to plant, and usually the crops gets ready for harvest from April to May.

Mid-November to the beginning of December is recommended areas with heavy frost and, with harvesting taking place from April to May.

It is common to plant from January to March in frost-free areas so that the growing season extends through winter. Cold spells during winter can be a risk, depending on the climate of the specific area. In very hot areas, planting should be avoided from November to middle of February as storage root formation is reduced by high temperatures.

Spacing Optimum plant density depends on cultivar, but is usually around 40 000 plants per hectare. Rows may vary from 1 to 1,25 m apart; in-row spacing it is usually 25 to 30 cm. Seeding rate The number of cuttings required to plant 1 ha varies between 30 000 and 60 000, depending on the specific spacing used.

Spacing

Optimum plant density depends on cultivar, but is usually around 40 000 plants per hectare. Rows may vary from 1 to 1,25 m apart; in-row spacing it is usually 25 to 30 cm.

Seeding Rate

The number of cuttings required to plant 1 ha varies between 30 000 and 60 000, depending on the specific spacing used.

Fertilisation

The recommended fertiliser rate for sweet potato production is based on crop removal figures. Research has shown that this recommendation will produce high yields when used in conjunction with yearly soil nutrient testing and petiole sap nutrient monitoring. Estimated crop removal in kg per ha is:

- 100 kg Nitrogen (N)

- 90 kg Phosphorus (P)

- 200 kg Potassium (K)

- 200 kg Calcium (Ca)

All the phosphorus may be applied in the basal along with 50 kg of N and 50 kg of K. The remaining 50 kg N and 150 kg K should be divided into two side-dressings at 4 to 6 weeks and at 10 to 12 weeks from planting. Some calcium will be supplied by the lime or dolomite used to adjust the soil pH, and any additional calcium may be applied in the basal as gypsum. Petiole sap nutrient monitoring is advisable so that the desired nutrient levels for different growth phases can be checked. Any trace element defi ciency would be detected by regular petiole testing, but generally, two foliar applications around the time of side-dressing should maintain adequate levels. Sprays should include zinc, copper, manganese, iron and boron.

The following table shows the optimum ranges for the major nutrients in petiole sap:

Fertiliser	Early running (0 – 10 weeks)	Mid growth (10 – 15 weeks)	Late growth (15 – 20 weeks)
Nitrate ppm	2 000 – 3 000	1 000 – 2 000	500 - 1000
Phosphate ppm	100 - 200	100 - 200	100 - 200
Potassium ppm	3 000 – 4 500	3 000 – 4 500	2 500 – 4 000
Calcium ppm	300 - 700	300 - 700	300 - 700
Magnesium ppm	300 - 700	300 - 700	300 - 700

Fertiliser Application

The recommended rates of side-dressing fertiliser should be calculated on the crop area (e.g.: 20 rows, 50 m long at 2 m spacing = 2 000 m² or 0,2 ha). If using drip tape, this fertiliser needs to be injected through the lines. If watering with sprinklers, then the fertiliser can be either injected or applied in the solid form and irrigated into the beds.

Irrigation

Sweet potato responds well to increasing moisture, but is considered a drought-tolerant crop because it is deep rooted and capable of developing storage roots under very dry conditions.

Requirements for water vary with soil type but can be generally estimated as 18 to 20 mm per week early in the season, 40 to 45 mm per week during the middle part of the season when storage roots are enlarging rapidly and a reduction to about 20 mm late in the season. Excessive moisture early in the season delays storage root development and enlargement; late in the season, it induces cracking and/or rotting of roots.

Irrigation and Scheduling

The best way to maintain the desired moisture content in the hills is to monitor the soil moisture with tensiometers. A shallow tube at 15 to 20 cm into the hill will indicate the timing of irrigation, and a deeper tube just below the base of the hill at 40 to 50 cm will determine the length of irrigation. Water requirement will vary with soil type and increase as storage roots develop within the hills. It is desirable to maintain moisture content in the hills at or near field capacity. Both tensiometers should remain within the 10 to 20 kPa range on sandy soils. This is especially important from 10 weeks onward as roots have been initiated and are starting to fill out. This is also the period of increased water use by the plants. Fluctuating soil moisture levels during this stage will reduce yield and cause cracking of roots.

Weed Control

Weeds may be a problem early in crop growth before vigorous vine growth covers the beds as plants become established. A number of control strategies may be used:

- After bed formation, irrigating should be applied to germinate any weed seed. Spraying with a knockdown herbicide before planting has been an effective method.

- Rotary finger cultivators are effective in removing small seedling weeds during early crop growth. Encourage vigorous early vine growth to smother weeds.

Pest Control

A fallow period should follow each crop to prevent build-up of soil-borne pests and diseases. Planting a green manure crop after harvest helps to suppress any regrowth and weeds as well as improving soil structure, and is essential for the long-term health of the soil.

Sweet Potato Weevil

This is the most serious pest of sweet potato. Adults are ant-like and lay eggs on stems and roots. The larvae burrow into the roots, making them unmarketable. They can pupate in the stems and be transferred in planting material. Once established in a crop, this pest is difficult to control. Research has shown that a pre-plant treatment of cuttings with chloropyrifos combined with foliar applications of chloropyrifos at 5 and 10 weeks from planting provides signifi cant control. Planting material collected from an infected crop would require insecticide dipping before planting. Destroying all crop residues after harvest and crop rotations are the best ways to keep weevil numbers down.

Giant Termite

Termites can be a major problem, especially on newly cleared ground where the activity of established colonies has not been identified. Avoiding known termite-infested areas may be successful in the short term. Aggregation techniques to locate and concentrate termite activity followed by a baiting programme is the best way to clear future planting areas of this pest.

Other Pests

Leaf-feeding caterpillars may cause problems if infestation is severe enough to cause significant leaf reduction. At the start of the wet season, hungry magpie geese can cause serious damage by trampling crops and eating the roots. Black-footed tree rats are also a problem.

Disease Control

Mycoplasma (Little Leaf Disease)

Infected plants have small, pale-yellow, stunted leaves and stems. The infection is spread by leafhoppers and if plants are infected while young, yields are greatly reduced. Control is by regular monitoring for symptoms and the removal and destruction of infected plants. Fungal disease Soil-borne fungal diseases can infect the roots but are not a large problem on well-drained, sandy soils. Any organic matter added to the soil should be well decomposed before planting.

Viruses

Feathery mottle virus has been detected in various sweet potato growing areas but research has shown that the infection had no signifi cant effect on yield (1993 Virus effect on yields). In other major production areas of Australia, severe infection has caused yield reduction and distorted roots. Symptoms are often not visible on infected plants, and laboratory testing is required to confi rm any infection. The virus is spread by insect vectors and by infected planting material. If sweet potato is to be grown over an extended period, then new virus-free material should be obtained from the virus-free programme every few years.

Other Cultivation Practices

Bedding Seed Roots

Before bedding sweet potatoes for plant production, examining roots carefully and discarding diseased, mutated and bruised roots should be the first step. Treatment of seed potatoes with a recommended fungicide dip immediately before bedding is needed. Dipping will help control surface infestations of black rot, scurf and root rot organisms. Washing seed potatoes before fungicide treatment allows for more efficient removal of all diseased potatoes and removes dirt that reduces the effectiveness of the chemicals. Seed potatoes should not be washed unless they are treated in a fungicide dip before bedding.

The fertiliser should be mixed with the bedding material. Warm beds to 26°C prior to bedding, then lower the temperature to 21 to 24°C once sprouting begins. Treated roots can be placed in the bed so that they are not in contact with each other. About 1 m² of bed is needed per volume of seed potatoes. Mesh wire prevents roots from being pulled along with slips. Roots must also be separated according to size to get an even depth of covering and uniform sprouting.

After bedding, the roots, water should be sprinkled over the bed to slightly moisten the soil (not soggy wet). Tar paper or plastic can be placed directly over the plant bed surface. When the slips push the covering up about 5 cm, the covering material should be removed. Watering the beds is important to keep the soil moist. The beds should be kept covered with sash or fi lm plastic until the plants begin to emerge. Ventilating during the day is necessary to control air temperature in the beds after the emergence of the plants. Air temperature in the beds should be kept under 32°C to produce good-quality plants. The plants should be pulled when they are about 20 cm tall. They should have at least five leaves, stocky stems and a healthy root system. This type of plant is best for mechanical transplanting.

Harvesting

Sweet potatoes, which bruise easily, are harvested by hand with mechanical aids. Vines are mechanically removed. Large tractor-drawn platforms that have a digger chain running in the centre are frequently used to lift the sweet potatoes from the ground. The sweet potatoes are then carefully removed from the chain by hand and placed either in wooden boxes holding 18 to 23 kg or in a bin that measures by 1.2 by 1.2 m and holds 454 kg.

Root maturity can vary between varieties, and root development is slower during cooler weather. Growers need to monitor the development of roots with regular checks of root size after 18 weeks. Marketable grades of roots are between 0.25 and 1 kg. If harvested at the correct time, around 60 to 70% of total roots should be within this grade. If grown during the dry season, most varieties should be ready for digging up at about 20 to 22 weeks from planting. If left too long in the ground, the roots can become oversized and unmarketable.

Harvesting Methods

Harvesting sweet potato can be very labour intensive, and requires suitable equipment for commercial production. Before harvesting, most of the top growth needs to be removed or it will become entwined in the digging machine. Vine removal is best done with a swinging pulveriser where the fl ails are shaped to the contour of the bed. This will chop the vine into pieces and leave the hills bare. A standard slasher or pulveriser can be used, but will not remove material between the rows. Chopping into the top of the hill should be avoided at all costs as this may damage the roots. Following this, any remaining vines can be cut on both sides of the hill with large, sharp coulters mounted on a tool bar. This vine removal should be done a week before digging to toughen the skin of the roots.

Roots are lifted from the soil using a single row potato digger. To avoid digger damage, this should be done while the hills are still moist so that some soil travels up the digger bars with the roots. The digger elevator should be moving only slightly faster than ground speed.

The dug roots are then manually collected into bulk bins and transported to the shed. The harvested crop must be kept away from lengthy exposure to the sun, and skin damage will be less if the roots are kept wet during handling.

Research trials have shown that 20 to 40 t/ha of marketable roots is achievable, depending on variety and management.

Root Curing

Sweet potatoes to be stored for later marketing or for seed stock must be cured immediately after harvesting to minimise storage losses. Curing involves controlling temperatures and relative humidity and providing ventilation for seven to ten days. Curing is a wound-healing process which occurs most rapidly at 26 to 32°C, a relative humidity of 85 to 90% and good ventilation to remove carbon dioxide from the curing area. Wounds and bruises heal and a protective cork layer develops over the entire root surface. In addition, suberin, a waxy material, is deposited. The cork layer and suberin act as a barrier to decay-causing organisms and to moisture loss during storage.

This process involves the forced hot air treatment of roots at 30°C with 90% relative humidity for between 4 to 6 days. This must be done immediately after harvest, and will result in the formation of a wound skin, which heals any mechanical damage suffered during harvesting. Post-harvest rot infections are minimised and excessive moisture loss prevented. Curing can also improve eating quality by increasing sweetness.

Root curing is not a standard commercial practice, but is worth considering if roots need to be stored for a prolonged period. Subsequently, harvested roots are placed in buildings to cure (30-35°C, 90% RH) and then stored (10-15°C; 85-90% RH) until needed for the market. Curing promotes wound healing and provides a barrier to prevent bacteria and fungi from entering damage that results during harvesting and handling. Properly cured roots will store for 12 months or longer with 15 to 25% losses under the best conditions.

Temperature must not drop below 12°C in order to prevent physiological cold damage to which sweet potatoes are particularly susceptible. Relative humidity should remain between 80 and 90% to prevent dehydration as the living storage roots continue to respire. As they are needed for marketing, roots are removed from storage rooms, processed through a mechanical washer/grader and packed into boxes of about 15 kg. Wash water may contain chlorine or other approved fungicides to reduce infection of wounds generated by the grading procedure.

Beetroot

Beetroot is a close relative of Swiss chard and sugar beet, and has many health benefits. The young leaves are tasty, a good source of vitamin A and can be prepared in the same way as spinach. The beets are rich in vitamin C.

Beetroot is a cool weather crop, but the hybrid (F1) cultivars available for summer production offer many advantages. The seed is expensive, but these beets are worth growing because they are better quality, more adaptable to extreme high temperatures and so are more uniform in shape, produce greater yields and have better internal colour. Hybrids also taste better, especially out of season.

Climate

- Beetroot is usually grown in cool regions or during the cooler seasons in warm areas.
- The growing period varies from 8 to 11 weeks in favourable climatic conditions.
- In hot weather the quality is adversely affected, which is shown by the alternate white and red rings when the beets are sliced.

- High-quality beets are characterised by a high sugar content and dark internal colour.

- The best planting times for beetroot are spring and autumn.

- The optimum temperature for growth is between 15°C and 20°C.

- Beets are not particularly sensitive to heat, as long as there is enough moisture in the soil. Although tolerant to cold, they grow extremely slowly in winter.

- Leaves may be damaged and growth retarded if there is frost before harvesting.

- Cold weather might delay maturity and the tops tend to be smaller.

- Direct sowing can result in good germination at temperatures between 6°C and 24° C.

- On hot sunny days, high temperatures that develop at, or just below, the soil surface might injure young plants badly, or kill them.

- High temperatures for long periods may not only retard growth and depress yield, but could also cause an undesirable strong flavour, concentric rings and a coarse texture.

Soil Requirements

- Sandy to deep, well-drained sandy loam or silt loam, high in organic matter, is recommended. Cloddy, stony, poor or very shallow soils are not suitable.

- Uniform soil moisture is essential for good quality.

- If the soil is compacted or the clay content is very high, roots are likely to be deformed and to develop a tough texture that reduces quality.

- Crops thrive in deep, rich sandy loam, with a pH of between 6 and 6.5 (but not below).

Raised Beds

- Raised beds can increase the effective depth of soil, allowing it to drain better, concentrate topsoil around the root zone, and provide more oxygen for healthy root development.

- Aeration is better, and disease, infection and the incidence of damping off are all reduced.

- Raised beds are truly beneficial if soil is heavy and/or poorly drained. Harvesting is also easier.

- Raised beds should be 1 m to 1.2 m wide with 50 cm between them.

- If you're making them by hand, mark the area with twine, then use a spade and a rake to make the beds.

- Large-scale farmers obviously use special equipment to make beds.

- If you have 1.2 m wide beds, 6 rows or furrows that are 2 cm deep would be good spacing. Start the furrows 10 cm from the side of the seedbed and allow 20 cm between rows.

- Sow the seed 2 cm to 3 cm apart and cover the furrows firmly with the soil from the furrows.

Direct Sowing

- It is essential that farmers buy quality seed that has a good germination percentage.

- It is very important to establish a fine, level seedbed when sowing the seed and to irrigate lightly a day before sowing.

- If done by hand, try to sow the seed evenly in the furrow about 3 cm to 4 cm apart.

- Do not sow too densely – that makes later thinning of the plants uncomfortable.

- Thin plants to 5 cm to 9 cm apart in the rows, depending on the size of beets needed for a specific market.

- If possible, sow seeds when the weather is cloudy.

Transplanting

More than 90% of beetroot producers sow the seed directly in the soil, but seed can also be sown in seedbeds and transplanted.

Seed trays or other containers can also be used to raise seedlings but this is expensive because of the high cost (about 450 000 plants are needed to establish 1 ha).

Mulching

Mulching can protect emerging seed from burning and keep the top soil layer moist and cool. Mulching materials include straw, corn cobs, sawdust, sunflower seeds, peanut shells, grass, grass clippings, newspapers and household waste.

- Good mulch must be inexpensive, available and easy to handle. It must also be stable, so that it will not easily wash or blow away.

- Remember that it's the temperature of the soil, not of the air, that controls seed germination so it is best to wait for soil temperature to rise before sowing seed.

- In summer, mulch has a cooling effect on the root system.

- A good layer of mulch can reduce evaporation from the soil surface by as much as 70%.

Fertilising

- A soil analysis or test is the most accurate guide to fertiliser requirements.

- Recommended soil sampling procedures should be followed in order to estimate fertiliser needs, and good management practices are very important if optimum fertiliser responses to beets are to be realised.

- Top or side dressings of nitrogen should be applied at about 100 kg/ha or (10 g/m^2) at the three-leaf stage, about 3 weeks after emergence, and 100 kg/ha 3 weeks later.

- Potassium levels should be kept fairly high. The second top dressing can be1:0:1 or potassium nitrate if K levels are low.

- Beetroot prefers well-drained soil, well-supplied with lime and potash.

- Heavy soils usually are not so likely to run short of potash.

- A lack of phosphorus or nitrogen will stunt growth and produce a deep red colour.

- When grown extensively under irrigation, beets can tolerate high salt concentrations.

- Beetroot are sensitive to high acidity and low boron levels.

- They are, in fact, a good indicator for boron deficiency: blackened areas and cracked roots are usually signs and, when cooked, there are black spots throughout the tissue and the beets taste bitter.

Irrigating

- Always irrigate carefully and, early in the season, take care not to irrigate too much.

- Waterlogging can turn leaves red and plants may stop growing for a while.

- As a general guide, apply 300 mm to 350 mm water throughout the growing season, starting off with 20 mm in the first week and 40 mm every week thereafter.

- Irrigation is especially important in the early stages of plant development and during root development.

- When sowing beetroot, keep the soil damp, lightly irrigating often to keep the surface cool, especially in warm weather.

- The growth points of emerging seed are very sensitive to hot soil conditions, so during long spells of hot, sunny weather, give about 8 mm water per day.

- On cold winter days, about 2 mm of water is needed.

- It is critical to irrigate the field in the last half of the growing season.

- Water shortages at this time could have the greatest negative influence on yields.

- During this period irrigate early in the day so that leaves can dry off and prevent diseases developing.

Harvesting

- Soil should be slightly moist before cutting or pulling beets.

- If the soil is too dry, roots may be difficult to clean and the rate of top breakage may be too high.

- For best flavour and tenderness, harvesting should begin when roots are 3 cm to 4 cm in diameter.

- Most beets grown commercially, however, are harvested when they are fully mature to obtain the highest yields.

- Handle beets carefully after harvesting to avoid damaging the roots.

- Damage reduces shelf life and increases the chances of decay and disease.

- Fresh market beets can be stored for 10 to 14 days, at 0°C and 98% to 100% relative humidity.

Carrot

Carrots are a cool season crop that is direct seeded in the spring for main season production or in the summer for fall and winter storage. They are in the family apiaceae, related to parsley, celery, parsnips and cilantro. They are a biennial, meaning they will produce a flowering head and seeds in their second year of growth. For most growers, carrots are not allowed to get to their seed-producing stage, as the carrot root is harvested in the first year.

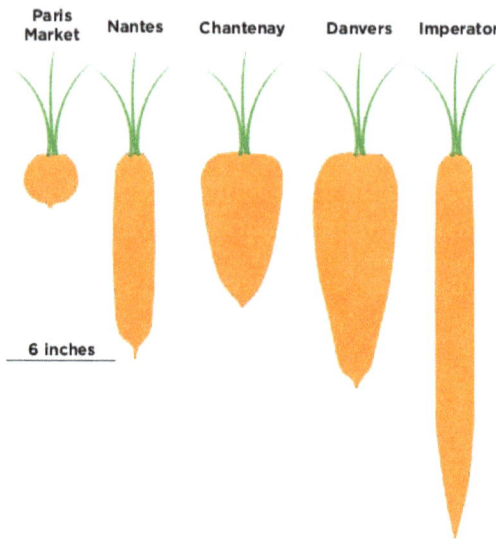

Common carrot root shapes

Carrot Types

Imperator	Most commonly grown type for bagged and baby carrots; long (8-10"), slender roots with small core and bright color.
Nantes	Mid-length (6-7") carrot with blunt tip and cylindrical form; known for excellent flavor and quality.
Danvers	Mid-length carrot with conical shape and thick form (2-2.5"); often used for processing, but can be used for fresh marketing when harvested young.
Chantenay	Primarily a processing carrot due to high yields and coarse texture; short to mid-length (4.5-5.5") with large shoulder and conical shape.
Paris Market	Palm-sized, round carrot with sweet flavor; commonly grown in less than ideal soil conditions when longer varieties struggle to perform.

Soil Preparation

Carrots prefer well-drained sandy loam or muck soils with a neutral or slightly acidic pH (5.5-7.0). Sandy soils will produce straighter, more uniform carrots, while heavy soils with rocks or debris will result in shorter, forked roots.

Pre-season preparation is essential for maximum production. Cover cropping with a grass-legume mix will help manage weed pressure, build organic matter and maintain good soil structure. Commonly used are a pea and oats blend for a spring carrot crop or a rye and vetch blend for fall carrot crop.

Properly weeded organic carrots at the north farm

Chisel plowing, keyline plowing or broadforking to a depth of 12-20" will assist in creating deep channels for carrot roots to grow. Breaking up compacted soils will provide the best opportunities for straight taproot production. Shanks can be aligned directly with where carrot rows will be seeded, ensuring a clear path of growth. A fine seedbed is important when planting carrots, as the small seeds need to have good soil contact and water absorption to ensure proper germination. Most growers will use a rototiller, while others will use a field cultivator with a crumbling roller, to create a fine seedbed in preparation for seeding.

Most growers will also incorporate a bare fallow period to ensure adequate weed control before planting a field into carrots. This can take place during the previous summer, prior to establishing a fall cover crop, or can be a few weeks of stale bedding prior to planting. Shallow cultivation will eliminate emerging weeds without lifting buried weed seeds. Flex tine weeders, basket weeders, and flame weeders are commonly used in the stale bedding process.

Fertility Requirements

Carrots are generally good nutrient scavengers, due to their deep taproot. All fertility applications should follow recommendations based on soil testing.

Fertility is usually provided through two applications: 50% pre-planting, and 50% side dressed. Sandy soils generally require higher levels of fertilization than heavy soils. 100-150 pounds of Nitrogen per acre is generally applied for adequate growth. Soil phosphorus levels should be greater than 30 ppm (bicarbonate extraction method) or 70 ppm (Bray extraction method). Exchangeable potassium levels between 100-200ppm generally do not require additional fertilization.

Crop Establishment

Carrots germinate in soil temperatures of 40 degrees Fahrenheit and warmer, while root and leaf growth occurs best between 60-70 degrees F. Carrots are direct seeded at 12-15 seeds per row foot, with 15-24" between rows on flat ground or in raised beds of 4" or more. Multi-line bed systems are also often used with 6-8 lines in 3-4 rows per bed. Two to four pounds of raw seed are need to seed an acre of fresh market carrots, which equates to 0.9-1.3 million seeds per acre, dependent on variety. Precision seeders are valuable tools in carrot establishment to avoid the need to thin and ensure an adequate stand for heavy yields and easy cultivation. Commonly used precision seeders for small-scale growers include push models (Earthway, Jang or equivalent) or vacuum or belt seeders (MaterMacc, Stanhay or equivalent). Pelleted seed can be used to make precision seeding easier, though organic producers must make sure any seed coating is NOP-compliant.

Proper seed spacing results in uniform, straight and nicely sized carrots

Consistent moisture is essential for uniform emergence. Most growers utilize overhead irrigation systems to ensure adequate soil moisture at germination and ease of cultivation. Carrots are slow to germinate (7-10 days) and need consistent moisture during this time. Small growers can use row covers or other methods to reduce evaporation during the germination phase.

In Michigan, carrots need between 10-14 inches of water during the growing season, which can be supplied by irrigation or precipitation. Excessive water or moisture stress can cause cracking and deformities. Generally speaking, 1-1.5 inches of water per week during the growing season is adequate.

Weed Control

Much of a carrot crop's weed control must be taken care of pre-planting, though cultivation is part of any organic carrot production system. Pre-emergence weed control using flame-weeding technology is commonly used because growers can control weeds without disturbing the soils surface. Timing is key with this method and only close monitoring of conditions will ensure proper timing. A common method to estimate is to seed a sample of beets at the end of a carrot row. When the beets emerge (one to three days before the carrots), the field can be flame-weeded.

Post emergence weeding can be a challenge for the organic grower. Carrots have little competitive advantage due to their small size and slow growth. Early management of between-row weeds can be accomplished with shallow cultivation using basket weeders, tine weeders, or tender hoes/sweeps. When the crop has reached a larger size, finger weeders and more aggressive shovels and sweeps can be used to control between- and in-row weeds and cover carrot shoulders. Depending on weed pressure and control methods, hand weeding may be required.

Harvest, Handling and Storage

Fresh market, processing, and storage carrots can be harvested throughout the season, though cool season harvest will reduce losses. Small growers may choose to hand-dig carrots using digging forks or an undercutter bar. Growers on larger acreages will use single or multi-row digging equipment that lifts, tops and bins carrots.

Carrots stored at 32 degrees Fahrenheit and 98-100 percent humidity will last 7-9 months, depending on variety. Prompt post-harvest cooling will aid in long-term storage. Carrots are generally stored in sealed plastic or wooden macro bins, harvest lugs or plastic bags to conserve moisture. Most small producers store carrots dirty, though many growers have had successes with storing cleaned carrots. Carrots are generally cleaned using barrel washing equipment or automated brush wash lines on small scales. Washing prior to storage often reduces staining of the roots, but can occasionally lead to decay due to bruises and nicks that occur during the washing process. The use of sanitizers in wash water can reduce the presence of decay-causing organisms, potentially increasing storage life.

Carrots can be a profitable crop for the small farm, not only during the growing season, but also throughout the winter months, when storage crops help maintain consistent relationships with customers. Proper management can lead to heavy yields of this universally loved vegetable.

Radish

Radishes are a hardy, easy-to-grow root vegetable that can be planted multiple times in a growing season. Radish seeds can be planted in both the spring and the fall, but growing should be suspended in the height of summer, when temperatures are typically too hot. (Hot temperatures may cause radishes to bolt, making them essentially useless.) Otherwise, radishes are one of the easiest vegetables to grow.

Planting

Ways to Plant Radishes

- Like carrots, radish plants are primarily grown for their roots. The roots will not grow well in compacted soil, so be sure to till your garden bed and remove any rocks before planting. If your soil is clayey, mix in some sand to loosen it and improve drainage.

- Incorporate a few inches of aged compost or all-purpose fertilizer into the planting site as soon as the soil is workable. Radishes do best in soil that's rich in organic matter.

- For a spring planting, sow seeds 4-6 weeks before the average date of last frost.

- Directly sow seeds outdoors ½ inch to an inch deep and one inch apart in rows 12 inches apart.

- Plant in a sunny spot. If they are planted in too much shade—or even where neighboring vegetable plants shade them—they put all their energy into producing larger leaves.

- Practice three-year crop rotation. In other words, only plant radishes in the same spot every third year. This will help prevent diseases from affecting your crop.

- Plant another round of seeds every 10 days or so—while weather is still cool—for a continuous harvest of radishes in the late spring and early summer.

- Plan on a fall planting. You can plant radishes later than any other root crop in late summer or early fall and still get a harvest. Sow seeds 4–6 weeks before the first fall frost.

Care

- Thin radishes to about two inches apart when the plants are a week old. Crowded plants do not grow well.

- Consistent, even moisture is key. Keep soil evenly moist but not waterlogged. A drip irrigation system is a great way to achieve this.

- Putting a thin layer of mulch around the radishes can help retain moisture in dry conditions.

Pests/Diseases

- Cabbage Root Maggot

- Clubroot

- Weeds: Weeds will quickly crowd out radishes, so keep the bed weed-free.

Harvest/Storage

Harvesting Radishes

- Radishes will be ready to harvest quite rapidly, as soon as three weeks after planting for some varieties.

- For most varieties, harvest when roots are approximately 1 inch in diameter at the soil surface. Pull one out and test it before harvesting the rest!

- Do not leave radishes in the ground long after their mature stage; their condition will deteriorate quickly.

- Cut the tops and the thin root tail off, wash the radishes, and dry them thoroughly. Store in plastic bags in the refrigerator.

- Radish greens can be stored separately for up to three days.

Turnip

The turnip or white turnip is a root vegetable generally grown in temperate climates, tropical and subtropical regions of India for its white, bulbous taproot. The most common type of turnip is coloured in white. This root vegetable belongs to the family of "Brassicaceae" and genus of "Brassica". The small and tender roots are used for culinary purpose where as bigger size turnips are used as animal feed.

Growth Habits

Turnip is a member of the mustard family and is therefore related to cabbage and cauliflower. Turnip is a biennial which generally forms seed the second year or even late in the fall in the first year if planted early in the spring. During the first or seeding year 8 to 12 erect leaves, 12 to 14 in. tall with leaf blades 3 to 5 in. wide are produced per plant. Turnip leaves are usually light green, thin and sparsely pubescent (hairy). In addition, a white-fleshed, large global or tapered root develops at the base of the leaf petioles. The storage root varies in size but usually is 3 to 4 in. wide and 6 to 8 in. long. The storage root consists mainly of the hypocotyl, the plant part that lies between the true root and the first seedling leaves (cotyledons). The storage mot generally has little or no neck and a distinct taproot. The storage root can overwinter in areas of mild winter or with adequate snow cover for insulation and produce 8 to 10 leaves from the crown in a broad, low-spreading growth habit the following spring. Branched flowering stems 12 to 36 in. tall are also produced. The flowers are clustered at the top of the raceme and are usually raised above the terminal buds. Turnip flowers are small and have four light-yellow petals.

Environment Requirements

Climate

Brassicas are both cold-hardy and drought-tolerant. They can be planted late-even as a second crop-and provide high-quality grazing late in the fall. Turnip planted in July will provide grazing from September to November. The most vigorous root growth takes place during periods of low temperature (40 to 60° F) in the fall. The leaves maintain their nutritional quality even after repeated exposure to frost.

Soil

Like other Brassicas, turnip grows best in a moderately deep loam, fertile and slightly acid soil. Turnip does not do well in soils that are of high clay texture, wet or poorly drained. For good root growth turnip needs a loose, well aerated soil.

Cultural Practices

Seedbed Preparation

Turnip seed is small and it is essential that it be seeded into a fine, firm seedbed with adequate moisture for germination. Plow and disk or harrow to produce a seedbed that is fine, firm and free of weeds and clods.

Turnip, like other Brassicas, can also be seeded into a sod or into stubble of another crop with minimum tillage. When seeding into sod, it should be suppressed or killed, as the young Brassica seedlings cannot compete with established grasses. To kill sod, apply 2 qt/acre of Roundup at least three days prior to seeding. A 0.5 qt/acre rate of Roundup can be used in 3 to 10 gal water/acre to suppress sod or to prepare a field of wheat stubble for seeding with turnip. Once established, turnip will compete with most weeds.

The advantages of direct drilling turnip into sod include fewer crop losses due to insect pests, such as the flea beetle, and less soil erosion on sloping sites where pastures are often located. A field of turnip established in sod gives animals a firm footing in all kinds of weather. It also allows the original sod to grow again the following spring if it has only been suppressed.

Seeding Dates

Turnip seed does not germinate well in cold soil. Turnip should not be planted until the soil temperature is at least 50° F or at corn planting time. The crop can be planted any time during the summer until about 70 days before a killing frost. Plantings after these dates may not have sufficient time to produce good forage growth.

Method and Rate of Seeding

Turnip seed can be planted in 6 to 8 in. rows at a rate of 1.5 to 2.5 lb/acre with a minimum-till drill when sod seeding. In conventionally prepared seedbeds, the crop can be seeded with a forage crop seeder or broadcast followed by cultipacking. The seed should not be covered with more than 1/2 in. of soil. A plant population of 5 to 6 per sq. ft. is desirable.

Fertility and Lime Requirements

Good soil fertility is very important for good yields. Soil tests should be taken to assure proper fertilization. Lime acid soils to pH 6.0. Fertilizers should be applied at the time of seeding or within 3 days of seeding to give the crop a competitive edge on weeds. Apply 100 lb/acre nitrogen to soils containing 2 to 5% organic matter, 120 lb/acre if less than 2% organic matter and 60 to 80 lb/acre if more than 5% organic matter. Requirements for phosphorus and potassium are similar to those of a small grain. In Wisconsin and Minnesota, when soil tests are in the medium range, about 20 to 30 lb/acre of P_2O_5 and 120 lb/acre of K_2O should be applied. Fertilizer applications should be banded at least 2 in. to the side and below the seed or broadcast. Boron and sulfur may also be needed. If the soil tests "low" in boron, apply 1 lb boron/acre on sandy soils, and twice this amount on other soils. Apply 10 to 15 lb of S/acre if a soil sulfur test indicates a need for this element.

Variety Selection

Three forage turnip varieties are recommended for use in the Upper Midwest: Green Globe, and York Globe from New Zealand and Sirius turnip from Sweden. In Pennsylvania, Green Globe and York Globe yielded more than Sirius at 60 days after planting, but Green Globe reached its peak yield later than the other two. Sirius yields were more variable from year to year than Green Globe or York Globe. The tops and leaves of Sirius have less glucosinolate than the other two varieties.

Weed Control

Weeds are generally not a problem once the turnip crop is established. However, sod and annual weeds should be controlled chemically and/or culturally before planting. Sod can be suppressed or killed with Roundup, as described under Seedbed Preparation. If annual weeds are present at planting time, eliminate them with a burndown herbicide such as Gramoxone. Tillage before planting can be used for weed control on a conventional seedbed.

Diseases and their Control

Turnip crops may suffer from clubroot, root knot, leaf spot, white rust, scab, anthracnose, turnip mosaic virus and rhizoctonia rot. in some cases, diseases can lead to crop failure if rotation or other control measures are not used. Resistant varieties are available for some diseases. To prevent problems with diseases, Brassicas should not be grown on the same site more than two years in a row. If clubroot is a problem, rotation should be six years.

Insects and other Predators and their Control

Turnip crops are attacked by two different flea beetles, which eat holes in the cotyledons and first leaves, chew stems and cause extensive plant loss. The cabbage flea beetle and the striped flea beetle feed exclusively on Brassicas, including related weeds such as yellow rocket. Problems with these flea beetles are much greater when Brassicas are grown under conventional tillage. Both flea beetles can be controlled with insecticides applied to the soil at planting.

Turnip crops can also be damaged by infestations of the common turnip louse or aphid. This insect feeds on the undersides of the leaves and may be so close to the ground that it is difficult to reach with a dust or a spray. In cases of severe infestation, the outer leaves curl and turn yellow. Aphid-tolerant varieties such as 'Forage Star' can give some protection against this insect.

Harvesting

Turnip plants are ready for grazing or green-chop when the forage is about 12 in. tall (70 to 90 days after planting). It is best not to wait too long because fungal diseases may begin to cut yields approximately 110 days after planting. The pasture should be grazed for a short time and the livestock removed to allow the plants to regrow. If grazed to a 5 in. stubble, 1 to 4 grazing periods may occur, depending on planting date and growing conditions. Strip or block-grazing is desirable to insure complete grazing.

The forage quality of turnip is sufficiently high, especially in protein, that it should be considered similar to concentrate feeds, and precautions should be taken to prevent animal health problems. Livestock should not be hungry when put on pasture the first time so they do not gorge themselves. If the livestock are moving from a feed low in nutritional value, feed a high-quality diet for two to three weeks prior to grazing turnip, or feed turnip for 30 min/day for one week prior to heavier grazing. This will allow for the development of a rumen microbial population that is adequate to digest the high levels of protein in forage turnips. A lower quality hay should be made available (2 to 3 lb of dry roughage/head/day for sheep and 10 to 15 lb for cattle) to provide some fiber in the animals' diet.

Livestock should not feed on turnip during the breeding season or after the plants have begun to flower. Nitrate nitrogen toxicity can be a problem, especially if ruminants are allowed to graze on immature crops or if soil nitrogen levels are high. The risk may remain for a longer period of time in autumn than in summer. Dairy cows should not be fed more than 50 lb turnip/head/day and should not be milked immediately after feeding on turnip to avoid milk tainting.

Parsnip

A close relative of carrots, parsnips have a sweeter taste, with the best flavors expressed after exposed to a couple of weeks of cold temperatures or overwintered until the spring. They have most of the same cultural needs as carrots, but require a longer growing season. Parsnips are technically a biennial that sets seed in its second year of growth, but in production it is treated as an annual.

Parsnips go very well when roasted with other root vegetables, and can be used to replace carrots in many recipes, including in baked goods.

Site Selection

Parsnips do best when planted in full sun, but they will tolerate part shade. The best soil is well-drained with a pH of 6.0–6.8. Deep, loose, and fertile sandy loam and peat soils with good moisture-holding capacity grow the straightest and smoothest roots. Soils with rocks or hard clay clumps can cause forking or disfiguration of roots.

Planting in a raised bed can provide the necessary depth of tillage in the soil to ensure long roots. The soil should be worked to a depth of at least 2 feet.

Be sure that your soil does not have high levels of calcium, as excess calcium can be a factor in cavity spot. Too much nitrogen in the soil can encourage vegetative growth of the tops, rather than focusing energy on root development.

Direct Seeding – Raw Seed

In mild areas with short summers, sow in early to midspring, as soon as the soil can be worked –

parsnips are able to tolerate frosts. Parsnips germinate best when the soil temperature is 59– 77° F/15–25°C. If growing in areas with long growing seasons and hot summers, plant in early summer when there is still approximately 4 months until the first fall frost.

Sow seeds in a 2 inch band, about 1 inch apart, the equivalent of 20 seeds per foot, ½ inch deep, in rows 18–24 inches apart. This spacing should allow for 3–4 bands per bed. Parsnips may take as long as 2–3 weeks to germinate. Do not allow the soil to dry out prior to emergence.

Thin plants to 2–3 inches apart when they have 2–3 true leaves or are 6 inches tall. To avoid disturbing the roots of seedlings you want to retain, use scissors to clip thinned plants.

Direct Seeding – Pelleted Seed

If using pelleted seed, time your plantings the same as if you were using raw seed. Sow 1 pellet ever 2– 3 inches, ½ inch deep, in rows 18–24 inches apart. Thinning is not necessary.

Be even more persistent in keeping the soil moist consistently throughout the germination period. The initial waterings can sometimes supply only enough water to split or dissolve the pellet. If the soil dries out before the germination period is over, the seed may receive insufficient moisture of optimal germination. In this case, it may take longer than the usual 2–3 weeks for germination.

Weed Management

Keep the soil free of weeds throughout the entire growing season. Young plants do not compete well for resources, such as nutrient and light, with weeds, though older plants develop enough foliage to shade out weeds.

Prior to planting, it is best to weed or cultivate several times to decrease the number of weeds present in the bed. Flame weeding can be particularly effective, and can also be done just before the seeds germinate.

Weed several times after germination. Use only shallow cultivation as plants mature to prevent damaging the roots.

Disease

There are a number of leaf blights – including cercospora, septoria, alternaria, and xanthomonas – that can reduce the yield and quality of parsnips. These can be identified by leisons and curling of the leaves. If blights infect the crop early on in its maturity, the roots may not reach full marketable size. To reduce the risk of leaf blights, select varieties with resistances, monitor the amount of water on the leaves or only use drip irrigation, plant at the proper planting density, and avoid excess nitrogen that can cause overabundant top growth. Please consult your local Cooperative Extension Agent to positively identify the specific leaf blight.

Canker can be identified by the dark-colored lesions that form on the crown and shoulder of the roots. Keeping the crowns covered with soil for the full season and maintaining a crop rotation can help reduce incidence of canker.

Pests

The major insect pest of parsnips is the carrot rust fly. Larvae of the fly burrow into the roots, which makes them unmarketable. The best method of control is to maintain a 3–5 year crop rotation with any crop in the Apiaceae family, such as carrots, celery, and parsley. Row cover is another method of control, as it excludes insects.

Pre-harvest

Hill soil over the shoulders of the roots to prevent the occurrence of greening on the shoulders.

In the weeks nearing harvest, reduce irrigation or watering to prevent cracks developing in the roots.

Harvest

Begin harvesting in the fall, preferably after a period of cold weather. Cut or mow the tops and then fork or undermine, or use a root crop harvester for larger plantings. Be careful while harvesting roots, as parsnips tend to bruise easy; the light color of their skin allows any bruise to show readily. To increase the sugar content, leave in the ground throughout the winter. However, be sure to harvest all roots the next spring before the tops begin to regrow.

Balance harvests between spring and fall to allow for roots to be available for fresh eating in the fall, storage in the winter, and as an early spring harvest.

Storage

If you did not remove the tops prior to harvest, cut them before placing roots in storage. Hold roots, either washed or not, in perforated bags or bins at 32° F/0°C and 95% relative humidity. Parsnips may last in storage for 4–5 months. Do not store with vegetables or fruits that product ethylene, as over-ripening and rotting may occur.

Roots can also be stored in damp soil or sand in boxes or buckets.

References

- James D. Mauseth (2009). Botany: an introduction to plant biology. Jones & Bartlett Learning. pp. 145–. ISBN 978-0-7637-5345-0. Retrieved 28 September 2010

- Onion, seedproduction: hort.vt.edu, Retrieved 14 July 2018

- Garlic-cultivation: farmingindia.in, Retrieved 29 May 2018

- Vegetable-production-growing-beetroot-successfully: africanfarming.com, Retrieved 28 June 2018

- Linda Berg; Linda R. Berg (23 March 2007). Introductory Botany: Plants, People, and the Environment. Cengage Learning. pp. 112–. ISBN 978-0-534-46669-5. Retrieved 28 September2010

- Small-scale-organic-carrot-production: msue.anr.msu.edu, Retrieved 31 March 2018

- Radishes, plant: almanac.com, Retrieved 19 June 2018

Fruit Vegetables

Fruit vegetables are seed-bearing parts of plants that are used as vegetables in different cuisines. Some of the common types of fruit vegetables, such as tomatoes, eggplants, cucumbers, zucchinis, pumpkins, etc. have been carefully analyzed in this chapter.

Tomato

Tomato (Solanum lycopersicum) is a flowering plant of the nightshade family (Solanaceae), which is cultivated extensively for its edible fruits. Labelled as a vegetable for nutritional purposes, tomatoes are a good source of vitamin C and the phytochemical lycopene. The fruits are commonly eaten raw in salads, served as a cooked vegetable, used as an ingredient of various prepared dishes, and pickled. Additionally, a large percentage of the world's tomato crop is used for processing; products include canned tomatoes, tomato juice, ketchup, puree, paste, and "sun-dried" tomatoes or dehydrated pulp.

Planting

Planting Tomatoes

- If you're planting seeds you'll want to start your seeds indoors 6 to 8 weeks before the average last spring frost date.

- Select a site with full sun and well-drained soil. For northern regions, it is very important that your site receives at least 6 hours of daily sunlight. For southern regions, light afternoon shade will help tomatoes survive and thrive.

- Two weeks before transplanting seedlings outdoors, dig soil to about 1 foot deep and mix in aged manure or compost.

- Harden off transplants for a week before planting in the garden. Set transplants outdoors in the shade for a couple of hours the first day. Gradually increase the amount of time your plants are outside each day to include some direct sunlight.

- Transplant after last spring frost when the soil is warm.

- Place tomato stakes or cages in the soil at the time of planting. Staking keeps developing tomato fruit off the ground, while caging lets the plant hold itself upright.

- Some sort of support system is recommended, but sprawling can also produce fine crops if you have the space, and if the weather cooperates.

- Plant transplants about 2 feet apart.

- Pinch off a few of the lower branches on transplants, and plant the root ball deep enough so that the remaining lowest leaves are just above the surface of the soil.

- If your transplants are leggy you can remedy this by burying up to ⅔ of the plant including the lower sets of leaves. Tomato stems have the ability to grow roots from the buried stems.

- Water well to reduce shock to the roots.

Growing Tomatoes in Pots

- Use a large pot or container with drainage holes in the bottom.

- Use loose well-draining soil. We recommend a good potting mix with added organic matter.

- Plant one tomato plant per pot. Choose from bush or dwarf varieties. Many cherry tomatoes grow well in pots.

- Taller varieties may need to be staked.

- Place the pot in a sunny spot with 6 to 8 hours of full sun a day.

- Keep soil moist. Check daily and water extra during a heat wave.

Care

Tomato Plant Care

- Water generously the first few days.

- Water well throughout the growing season, about 2 inches per week during the summer. Water deeply for a strong root system.

- Water in the early morning. This gives plant the moisture it needs to make it through a hot day. Avoid watering late afternoon or evening.

- Mulch five weeks after transplanting to retain moisture and to control weeds. Mulch also keeps soil from splashing the lower tomato leaves.

- To help tomatoes through periods of drought, find some flat rocks and place one next to each plant. The rocks pull water up from under the ground and keep it from evaporating into the atmosphere.

- Side dress with fertilizer or compost every two weeks starting when tomatoes are about 1 inch in diameter.

- If using stakes, prune plants by pinching off suckers (side stems) so that only a couple of branches are growing from each plant. The suckers grow between the branch and the main stem.

- Tie growing stems to stakes with twine or soft string.

- As the plants grow, trim all the lower leaves off the bottom 12 inches of the stem.

- Practice crop rotation from year to year to prevent diseases that may have overwintered.

Pests/Diseases

Tomatoes are susceptible to insect pests, especially tomato hornworms and whiteflies.

- Aphids

- Flea beetles

- Tomato hornworm

- Whiteflies

- Blossom-end rot

- Late Blight is a fungal disease that can strike during any part of the growing season. It will cause grey, moldy spots on leaves and fruit which later turn brown. The disease is spread and supported by persistent damp weather. This disease will overwinter, so all infected plants should be destroyed.

- Mosaic virus creates distorted leaves and causes young growth to be narrow and twisted, and the leaves become mottled with yellow. Unfortunately, infected plants should be destroyed (but don't put them in your compost pile).

- Cracking: When fruit growth is too rapid, the skin will crack. This usually occurs due to uneven watering or uneven moisture from weather conditions (very rainy periods mixed with dry periods). Keep moisture levels constant with consistent watering and mulching.

Harvest/Storage

- Leave your tomatoes on the vine as long as possible. If any fall off before they appear ripe, place them in a paper bag with the stem up and store them in a cool, dark place.

- Never place tomatoes on a sunny windowsill to ripen; they may rot before they are ripe!

- The perfect tomato for picking will be firm and very red in color, regardless of size, with perhaps some yellow remaining around the stem. If you grow orange, yellow or any other color tomato wait for the tomato to turn the correct color.

- If your tomato plant still has fruit when the first hard frost threatens, pull up the entire plant and hang it upside down in the basement or garage. Pick tomatoes as they ripen.

- If temperatures start to drop and your tomatoes aren't ripening,

- Never refrigerate fresh tomatoes. Doing so spoils the flavor and texture that make up that garden tomato taste.

- To freeze, core fresh unblemished tomatoes and place them whole in freezer bags or containers. Seal, label, and freeze. The skins will slip off when they defrost.

- You can harvest seeds from some tomato varieties.

Eggplant

Eggplant, *Solanum melongena* L., is a popular vegetable crop grown in the subtropics and tropics. It is called brinjal in India and aubergine in Europe. The name "eggplant" derives from the shape of the fruit of some varieties, which are white and shaped similarly to chicken eggs.

Soil and Climatic Requirements

Eggplant is a warm-season crop and does not tolerate frost. A long growing season of 80 days is required for the transplanted crop. Optimal temperatures for eggplant production are 26°C days and 20°C nights.

Plant growth slows and pollination problems occur at temperatures below 17°C or above 35°C. Flowering is not affected by day length.

Cooler temperatures can reduce fruit set. Higher temperatures and high humidity levels also reduce yields. Eggplant can tolerate drought and excessive rainfall. It will not tolerate extended periods of saturated soil owing to the build-up of root-rotting pathogens.

Eggplant does well in a variety of soil textures. Previous crop residue must be stubble-disked to improve soil aeration and to adequately bury organic matter for decomposition. Eggplant grows best with a soil pH of 5.5 to 6.5.

Eggplant is usually grown in light or sandy loam soils that provide good drainage and favourable soil temperatures. Eggplant will root to a depth of 90 to 120 cm; therefore, sandy loam or silt loam soils free of physical barriers are better for proper plant growth and development.

Uses

This vegetable is quite diverse and more versatile, both in the garden and in the kitchen. Eggplant has chemicals that can cause digestive upset if eaten raw, so is usually cooked. It can be grilled, stuffed, roasted, served in soups and stews and on kebabs, and used in curries and stir-fries.

Human Health Benefits

Eggplant is nutritious, being low in calories, fat, sodium and is a non-starchy fruit that is cooked as a vegetable. It contains a large volume of water. It is good for balancing diets that are heavy in protein and starches. It is high in fibre and provides additional nutrients such as potassium, magnesium, folic acid, vitamin B6 and A.

Cultivation Practices

Soil Preparation

Well-drained, sandy loam soils are ideal for eggplant production. Poorly drained soils usually result in reduced functional root area, poor plant growth and low yields.

Site selection can be important if early eggplant production is required. For early production, select sites with a southern to southwestern exposure. Soil with a southern exposure receives more sunlight in the spring and therefore warms up more quickly.

Plan crop rotation so that eggplant is not planted after eggplant or other solanaceous crops such as tomato or pepper.

Good soil preparation is important for optimum eggplant production. If large quantities of plant debris are present, disk land several weeks before transplanting, then plow land, using a mouldboard plough. This will loosen the soil and bury old crop residue. Turn soils at least 20 cm deep.

Adequate soil preparation facilitates the growth and development of an extensive root system. Plants will then have a larger volume of soil from which to draw water and nutrients, reducing the chance of moisture and nutrient stress. Disking soil after turning can cause recompaction. If planting beds need to be made or smoothed prior to transplanting, use a rotary tiller or similar implement and maintain the same wheel patterns throughout subsequent operations.

Eggplant is intolerant of poorly drained soil, so it is usually helpful (especially on heavier soils or in low areas) to transplant eggplant on raised beds.

Planting

Eggplant crops are normally grown from transplants; however, a few growers use direct seeding. Desert growers plant spring transplants on southern-sloping beds that run from east to west. They use brush paper and wooden stakes to protect the crop from spring frosts. The butcher-type brown paper is held in place with wooden stakes placed every 60 cm along each row. The paper-stake structure is placed at an 80-degree angle to reflect sunlight downwards, warming the soil and young plants. The stakes must hold the paper securely, otherwise wind can cause it to vibrate and tear.

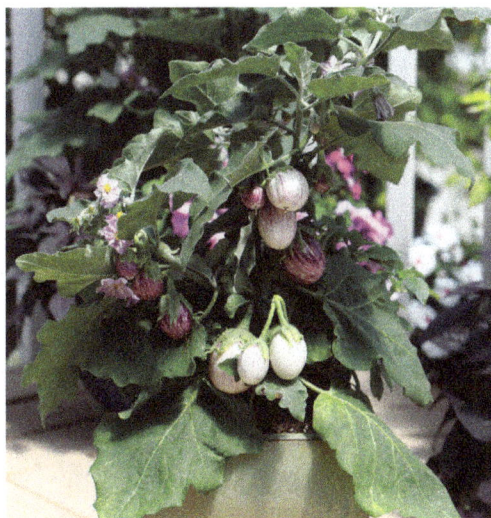

Clear, polyethylene mulch is also used on the spring crop. Some small-scale growers use mulch in combination with brush paper and stakes. Black-plastic mulch increases yields by controlling weeds, conserving moisture and warming the soil.

In-row spacing of eggplant is 30 to 60 cm. The crop can be grown, using a row width depending on the space needed by harvest workers. Growers usually plant 8 rows and skip 2 rows to make roadways for harvest operations. Growers are experimenting with a bed spacing of 45 to 70 cm in an effort to maximise sunlight penetration onto the fruit, improving fruit colour.

Some growers remove the lower leaves and flowers and stake the plants in an effort to reduce fruit rot that occurs when the fruit touches the soil.

Fertilisation

The nitrogen (N) requirement for eggplant is approximately 168 to 224 kg/ ha. Preplant fertilisers are usually broadcasted. A typical blend is 90 to 134 kg/ha each of phosphorus (P) and potassium (K) and 22 to 45 kg/ha of N. During the growing season, 2.3 to 4.5 kg of N is applied each week for the period of vegetal growth. At early flowering, 7 to11 kg of N is applied each week. During fruit enlargement, 5 to 7 kg of N is applied each week. The N is water run by most small growers.

Irrigation

Eggplant can be grown with furrow or drip irrigation. A crop of furrow-irrigated eggplant uses approximately 1850 m^3 of water. Some growers use black plastic mulch and drip tape to control weeds, moisture and soil temperature in spring plantings. Critical watering periods are at flowering, fruit set and enlargement. The volume of water applied, depends on the time of the year and stage of plant growth. Most of the water and nutrientabsorbing roots are in the top 45 cm of the soil. Irrigation should be managed to maintain good soil moisture in this root zone.

Pollination

Eggplant is self-fertile as its flower contains both male and female parts. Flowers are usually formed on opposite leaves. Flowering is considered day neutral. Eggplant is not well suited for greenhouse production because it will not set fruit in extremely high or low temperatures. Fruit abscission can result if day temperatures exceed 35°C. If night temperatures drop below 16°C, pollen deformity increases and less fruit is produced. Flowering and fruit setting begin 6 to 8 weeks after transplanting. Market size fruit is ready approximately 3 weeks after flowering.

Weed Control

Eggplant is slow to become established and cannot compete with aggressive weeds. Weeds also harbour damaging insects and diseases.

Weeds are controlled either by physical methods or chemical control. Physical methods, such as hand weeding, cultivation and mulching, are quite frequently used on small vegetable farms. Only shallow cultivation is necessary. Mulching with black plastic mulch effectively controls weeds and reduces labour needs. Natural organic mulches, such as rice straw, will conserve moisture and add organic matter to the soil.

Chemical weed control is especially popular in places where labour is expensive. Suitable herbicides include Lasso and Sencor (metribuzin).

Pest and Disease Control

Herbicides, insecticides and fungicides should always be used in compliance with the label instructions.

Insects

Many insect pests are attracted to eggplant. Spider mites (*Tetranychus* spp.), green peach aphids (*Myzus persicae*), lygus (*Lygus* spp.), flea beetles (Chrysomelidae) and wireworms (Elateridae) can be destructive to eggplant. Spider mites are especially harmful and should be treated as temperatures become warmer. Flea beetles are usually a problem only in young plants. Fields should be closely monitored during the flowering period as lygus will feed on flowers and cause flower drop. Root-knot nematodes (*Meloidogyne* spp.) can cause plants to wilt and leaves to yellow.

Diseases

Leaf spot and fruit rots caused by *Phomopsis vexans* are characterised by circular, brownish spots on fruit and leaves. Fruit rot may appear during postharvest transport even when symptoms are not evident at the time of harvest.

Control measures:

- Crop rotation with any other crop rather than solanaceous crop.
- The field should be cleaned as soon as the disease is detected in the field, i.e. the diseased fruit should be plucked and burnt.

Early blight caused by *Alternaria solani* can result in dieback known as collar rot in seedlings. Foliage can be affected at all growth stages, and fruit can drop owing to infection. This fungus is favoured at temperatures between 16 and 32°C. Stressed plants are more susceptible than healthy plants.

Control measures:

- Observing proper field sanitation
- Using certified disease-free seed
- Own seed should be water/heat treated

Anthracnose fruit rot from *Colletotrichum melongenae* causes sunken spots and lesions on the fruit surface. This fungus is favoured by temperatures between 13 and 35°C with optimum growth at 27°C and humidity at 93% or higher.

Control measures:

- Using resistant varieties, if available
- Using certified disease-free seeds
- Crop rotation
- Destroying infected crop residue.

Wilt caused by *Verticillium albo-atrum* affects the vascular system of a plant and results in stunted plant growth, yellow discolouration and eventually defoliation of the lower foliage and plant death. This fungus is favoured in temperatures between 13 and 30°C. Currently, there are no eggplant varieties available that are resistant to these soil-borne fungi.

Tobacco ring spot virus (TRSV) is characterised by yellowing foliage and plants dying off. Crop rotation can help to lessen the effects of this disease. The dagger nematode (*Xiphinema* spp.) is a known vector of TRSV. Postharvest losses of fruit can be caused by *Alternaria* spp. (black mould rot), *Botrytis* spp. (grey mould rot), *Rhizopus* spp. (hairy rot) and *Phomopsis* rots.

Harvesting and Handling

Harvest of eggplant usually starts 75 to 90 days after transplanting or 15 to 35 days after flowering expansion (anthesis). Fruit is harvested when it reaches market size, and the skin is glossy, but before seeds begin to enlarge significantly and mature. Varieties with elongated fruit take more time to ripen. Overmature eggplants become pithy and bitter. Fruit should be removed often to encourage continued fruit set. At market maturity, the fruit stem hardens and a sharp knife is needed to cut fruit from plants. The length of stem left on the plant can vary from 2.5 to 5.0 cm for American varieties and 2.5 to 7.5 cm for Asian varieties. Harvesting is done by cutting the stem rather than by pulling the fruit.

The fruit is dumped in a water bath for washing and cooling prior to packing. Fruit should be handled and packed carefully to avoid skin abrasions and puncturing. Some types of eggplant have skin that can be damaged easily. Careful harvesting and handling practices should be followed to avoid bruising and compression injuries. The fruit is packed by 18s and 24s into fibreboard containers.

Some growers cut plants to 45 cm, allowing them to grow out again for autumn harvest. This practice depends on current market prices and plant vigour.

Capsicum

Capsicum is variously called as green pepper, sweet pepper, bell pepper, etc. In shape and pungency it is different from chilli. It is fleshy, blocky, of various shapes, more like a bell and hence named bell pepper. Almost all the varieties of green pepper are very mild in pungency and some of them are non-pungent, and as such they can be used as stuffed vegetable.

Climate

It requires a similar climate like that of chilli and is also susceptible to frost. It prefers milder climate than chilli and 21 to 25°C is ideal for green pepper. Higher temperatures are detrimental to fruit set. High temperature and low relative humidity at the time of flowering increases the transpiration pull resulting in abscission of buds, flowers and small fruits. Moreover, higher night temperatures are found to be responsible for the higher capsicin (pungency) content in green pepper.

Soil

Although sweet pepper can be grown in almost all types of soils, well drained clay loam soil is considered ideal for its cultivation. It can withstand acidity to a certain extent. Levelled and raised beds have been found more suitable than sunken beds for its cultivation. On sandy loam soils, the crop can be successfully grown provided the manuring is done heavily and the crop is irrigated properly and timely. The most suitable pH range of soil for green pepper is 6 to 6.5.

Planting Requirement

The following procedures are followed to plant capsicum in the field.

Seedling Raising

Seedlings are first raised in the nursery beds and then transplanted in the main fields. Normally, 5-6 seed beds of size (300x60x15 cm) each are sufficient for one hectare cultivation. Seed should be sown in rows at 8 -10 cm apart to get healthy seedlings. The seeds should be dressed with Agrosan. Ceresan, Thiram or Captan 2 g per kg seed before sowing to prevent the occurrence of any seed-borne diseases. About 1 -2 kg seeds are required for one hectare cultivation depending on the cultivar. The seeds should be properly covered with a thin layer of soil manure mixture or any other media and irrigated with sprinkler to maintain optimum moisture till the seeds germinate.

Sowing Time

The sweet pepper is generally sown in August for the autumn-winter crop and in November for the spring - summer crop. In the hills of North Bengal sowing of seeds in the months of March - April (under cover) and September - October, is very successful for getting high yield. Plants sown in September and October take the longest period for development because of poor availability of light in winter.

Land Preparation

For planting the seedlings. the main field is thoroughly prepared by ploughing the land 5-6 time followed by smooth planking. Farmyard manure or compost is added after the first ploughing so that it is thoroughly mixed in the soil during subsequent ploughings. Then the field is brought to a clean and fine tilth.

Transplanting

The seedling having attained 4-5 leaves should be transplanted. The nursery beds should be irrigated before lifting of seedlings. The seedlings are transplanted in rows in the evening or during the cloudy day followed by irrigation. Generally, 50 to 60 days old seedlings are used for transplanting.

Spacing

The seedlings are transplanted in rows at a distance of 30 to 60 cm depending upon the area and the variety. Rows spaced at 90 cm and plants spaced at 40 to 45 cm are also fairly common.

Manures and Fertilizers

About 50 to 80 cartloads of farmyard manure, 30 to 55 kg of nitrogen in the form of ammonium sulphate or urea, 50 to 110 kg of phosphorus in the form of super phosphate and 75 to 100 kg of potash per hectare should be given depending upon the fertility status of the soil. The complete dose of farmyard manure should be applied in the soil at the time of first ploughing. Potassium and phosphate fertilizers should be mixed in the plant rows just before transplanting. The nitrogenous fertilizer is given two and half a month after transplanting.

Irrigation

The first irrigation is given just after transplanting and later the field should be irrigated as and when required. lrrigation is essential in arid and semi-arid regions.

Weed Control

Interculture operations are similar to that of chillies. Two weedings 30 and 60 days after transplanting lead to high yield in green pepper. Earthing of plants may also be done after 2 -3 weeks of transplanting. Earthing operation will also help in removing the weeds.

Insects and Diseases

The important insect. pests attacking capsicum are described here along with their suitable control measures.

Thrips

The adults of these tiny insects are slender yellow, active and pointed at both ends. The females have four extremely slender wings which have long fringe on their posterior margins. The male is similar to female except that it is smaller and lighter in colour. The minute insects lacerate the plant tissues and suck the sap from the leaves forming white blotches and curly leaves with stunted plant look. Consequently yield is reduced considerably.

Control:

It can be controlled by the spraying Malathion or Dimethoate. It may also be controlled by the spraying of 0.25% Nicotine sulphate.

Aphids

Aphids sometimes becomes serious on capsicum. They suck the cell sap from the leaves and petioles and cause considerable loss.

Control:

Complete control of aphids can be obtained by the application of Dimeton methyl (0.05 to 0.02%) or Monocrotophos (0.05 to 0.01%).

Mites

Mites of different genera have been found feeding on leaves of chilli and capsicum. These tiny spider like creatures may be found in large numbers on the underside of leaves, covered with fine webs. Both nymph and adults suck the cell sap and devitalise the plants.

Control:

It is reported that spraying of Phosaione (Zolone) 35 EC at 3 ml per litre can controi mites. -Spraying Dimethoate or Dicophol is very effective against mites.

Diseases

The important diseases affecting capsicum are described below.

Damping Off

This is a fungal disease which frequently occurs in nurseries. The seed may rot or the seedlings may be killed before they emerge from the soil. The stem of young seedlings may also be attacked after emergence showing water soaking and shriveling of stem which fall over and die. In a nursery, the disease may start in patches and in the course of 2 -4 days the entire lot may be destroyed. The disease is most damaging on moist soil with poor drainage conditions.

Control:

Partial sterilization of soil by burning trash in the surface helps in checking the disease. Providing better drainage by forming raised beds with free drainage all around helps in avoiding the disease. The seedlings may be protected by spraying with 0.5 to 1.0% Bordeaux mixture or any other effective copper oxychloride like Blitox or Fytolan.

Anthracnose

The foliage, stem and fruits are attacked by the fungus causing this disease. Disease areas on fruits develop as dark, round, sunken spots. Infected fruits drop off prematurely. Black minute spots develop on infected seeds. High humidity is favourable for the disease spread.

Control:

Treat the seeds before sowing with organo-mercurials such as Thiram (0.2%) or Brassico (0.2 per cent). Spray Difoltan (0.2%) or Ditbane M-45 or Blitox (0.4%) at 15 days interval.

Powdery Mildew

It is a serious disease of capsicum especially during summer. white talcum powder like growth appears on the leaf. Diseased leaves are shed and plants remain stunted.

Control:

The disease can be controlled by spraying Sulfex (0.2%) or Calixin (0.2%) at 15 days interval.

Bacterial wilt

This is a serious disease of capsicum affecting leaves as well as fruits. Characteristic symptom of bacterial wilt are rapid and complete wilting of normal grown up plants.

Control:

There is no chemical control for this disease. However, application of bleaching powder before planting 15 kg ha has been found very effective. The variety "Arka Gaurav" is known to be tolerant to this disease.

Leaf Curl Disease

This is an important viral disease of chilli and capsicum. Symptoms consists curling of leaves accompanied by puckering and blistering of interveinal areas and thickening mid swelling of the leaves.

Control:

The disease is transmitted by thrips and aphid. Thus it can be controlled by reducing the vector population. In the beginning, the plants showing infection should be uprooted.

Harvesting and Yield

Large sweet peppers usually are picked while they are still green in colour but fully grown when sold in the market. Some exotic varieties such as Pimiento and Paprika are harvested when fruits are GMk red ripe. However, the most favourable time for harvesting in Pimiento for seed production is between 50 -60 day old stage or when the fruit attain the bright to deep red colour. Sweet peppers are picked with an upward twist which leaves a piece of stem attached with the fruits. Young immature peppers are rather soft and yield readily to mild pressure of the fingers. Green fruits ready for harvest are relatively firm and crisp. The yield of capsicum varies depending upon variety and the method of cultivation. If proper care is taken during its growth, it may yield 10 to 12 tonnes of quality fruits per hectare.

Ripening and Storage

The bell type peppers are usually harvested and sold when they are of suitable market size and are still green. There is but a limited demand for the mature red specimens. Now-a-days different chemicals are used to accelerate ripening as well as inducing the fruit colouration. Spraying with Ethephon at 200 to 3200 ppm has been found effective to accelerate fruit colour development, fruit and leaf drop and leaf yellowing. Some varieties are harvested when they are completely red. Green peppers can be kept in good condition for at least 40 days at 00 C and at 95 to 98% relative humidity. The shrinkage of fruits stored under these conditions is only 4% in 40 days.

Seed Production

In seed to seed method of seed production, transplanting is done for commercial seed production. Optimum spacing may be followed to obtain high quality seeds. Capsicum is a cross pollinated crop, so it crosses easily with chillies and thus deteriorates fast in quality, if proper isolation distance is not maintained during seed production stages. The isolation distance between two cultivars of capsicum should be 200 meters for foundation seed and about 100 meters for certified seed production. Off-type plants are removed as soon as they are observed. The small leaved plants can be detected from the large leaved plants and should be rogued as per the requirement. The number of rouging should be 3 -4 depending upon the purity of the seed desired. Field inspection of seed crops should be done at least twice or thrice. The fruit should be picked when red ripe and cut and crushed or macerated by machine. Seed is to be washed thoroughly to make it free from pulp and skin. After washing, the seeds should be dried immediately in the sun. Picking of pods may be done according to climatic conditions. In case, there is no danger of rains at the maturity time, the pods may be picked in one lot but where there is some danger of rains picking may be done is 2 -3 installments.

Cucumber

Cucumber (*Cucumis sativus*) is a warm season crop that belongs to the family *Cucurbitaceae*, which is grown for its immature fruits.

There are different varieties of cucumbers, which vary in colour, size and different uses.

Cucumbers are great in salads, soups and dips. They could also be used to make pickles. They have very high water content and are very low in calories. If picked while young, cucumbers can add a nice taste to different foods.

Cucumbers are a good source of vitamin C.

Growth Stages

Vegetative Growth

Vegetative growth consists of 2 Stages:

- Stage I – Upright growth is the initial stage that starts when first true leaves emerge and it ends after 5-6 nodes.
- Stage II – Vining - starts after 6 nodes. Then, side shoots begin to emerge from leaf axils, while main leader continues to grow. Side shoots are also growing, causing the plant to flop over. Leaves are simple and develop at each node. Each flower/fruit is borne on its own stem attached to the main stem at a node.

Depending on variety and environmental conditions, flowers may begin developing at the first few nodes.

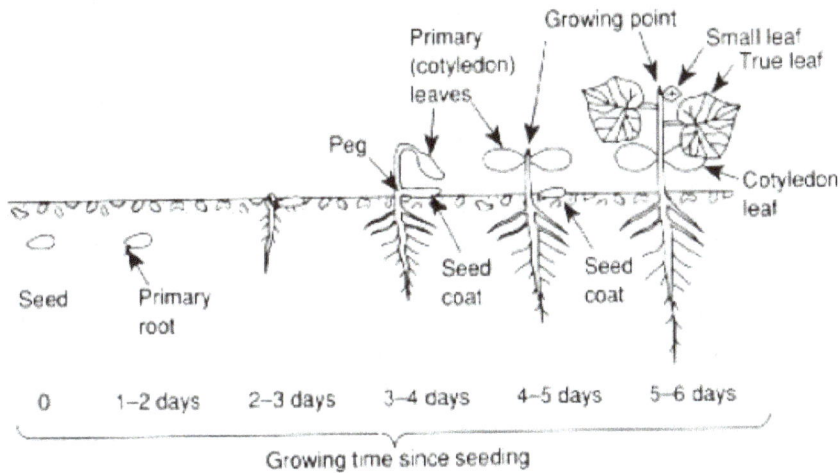

Figure: main development processes and organs developments during first 6 days from germination, under optimal conditions

Flowers and Fruits

Flower Types

There are different flower types:

- Staminate (male).

- Pistillate (female). Ovary located at base of the female flower.

- Hermaphrodite (both male and female).

Figure: Male flower

Figure: Female flower

Cucumbers are monoecious plants which have separate male and female flowers on the same plant. The male flowers appear first and female flowers shortly later. The female flowers have small immature fruit at the base of the flower and male flower do not have any. Pollen is transferred from male to female flower by bees or other insects. When pollinated properly, female flower develops into fruit. There are different types of cucumber hybrids such as gynoecious varieties that produce predominantly female flowers, and seeds of monoecious varieties are mixed with it for pollination. They are very productive when pollenizer is present.

Older cultivars, as well as many current cucumber cultivars, have a monoecious flowering habit, producing separate staminate and pistillate flowers on the same plant. Although the terminology is not botanically correct, staminate flowers are often referred to as male flowers and pistillate as female.

Monoecious cultivars first produce clusters of five male flowers at the leaf nodes on the main stem. Subsequently, the plant produces both male and female flowers.

Most current hybrids are gynoecious (all female flowers). Gynoecious hybrids are widely used because they are generally earlier and more productive. The term "all-female" is somewhat misleading as 5% of the flowers are male under most conditions. These modern F1 hybrids have several advantages. As they bear only female flowers the tiresome job of removing male flowers is unnecessary. They are also much more resistant to disease and rather more prolific. There are two drawbacks – the fruits tend to be shorter than the ordinary varieties and a higher temperature is required. Production of female flowers is naturally promoted by the short days, low temperatures and low light conditions of fall. Flower femaleness can be promoted by applying plant growth substances (PGRs) such as NAA (a type of auxin), and Ethephone (an ethylene promoter). If a purely female variety is grown, need to provide an appropriate pollinator.

In sensitive gynoecious cultivars, production of male flowers is promoted by long days, high temperatures and high light intensity typical to the summer season. Production of male flowers also increases with high fruit load and with stresses exerted on the plant. Maleness can be promoted by applying PGRs such as Gibberelins as well as by silver nitrate and AVG that act as ethylene suppressors.

Parthenocarpic Fruit

There are also cucumber hybrids that produce fruits without pollination. These varieties are called parthenocarpic varieties, resulting in fruits that are called 'seedless', although the fruit often contain soft, white seed coats. Such parthenocarpic fruit set also occurs naturally under the low-light, cool-night growing conditions, and short days of fall. Older plants can also produce 'super' ovaries which set fruit parthenocarpically.

Parthenocarpic varieties need to be isolated from standard varieties to prevent cross-pollination and development of fruits that do contain seeds, and may be deformed by greater growth in the pollinated area.

Greenhouse cucumbers are naturally parthenocarpic.

Male/Female Flowering Sequence

On a normal cucumber plant, the first 10 - 20 flowers are male, and for every female flower, which will produce the fruit, 10 - 20 male flowers are produced. Flowering set progressively at the nodes.

Developing fruit at the lower nodes may inhibit or delay fruit at subsequent nodes.

Size and shape of the cucumber fruits are related to number of seeds produced.

Pollination

Since each cucumber flower is open only one day, pollination is a critical aspect of cucumber production. One or more pollen grains are needed per seed, and insufficient seed development may result in fruit abortion, misshapen, curved or short (nubbin) fruit, or poor fruit set. Hence, 10 - 20 bee visits are necessary per flower at the only day the flower is receptive, for proper fruit shape and size. Therefore, it is important to bring hives into the field when about 25% of the plants are beginning to flower. Bringing in the bees earlier is unproductive because they may establish flight patterns to more abundant and attractive food sources such as legumes or wildflowers. Bringing them in later jeopardizes pollination of the first female flowers. It is important to take into consideration that bee activity is greatest during the morning to early afternoon, and that wet, cool conditions reduce bee activity and causes poor fruit set.

Cucumber varieties can cross pollinate with one another but not with squash, pumpkins, muskmelons, or watermelons.

Pollination in gynoecious ("all female flowers") crops is ensured by blending seed of a monoecious cultivar (pollenizer) with seed of the gynoecious hybrid. Typical ratios are 88% gynoecious, to 12% monoecious. Pollenizer seed is often dyed with a different color to distinguish it from that of the gynoecious hybrid. It is difficult to recognize pollenizer seedlings after emergence in the field. Removing 'different looking' seedlings during thinning may leave the field without the pollenizer.

Cucumbers types sorted by Final usage, Morphology and Culture Practice

Cucumber cultivars are usually classified according to their intended use as fresh market slicers, pickles, or greenhouse cucumbers. This classification includes several fruit characteristics such as shape, color, spine type (coarse or fine), spine color (white or black), fruit length/diameter ratio, skin thickness, and surface warts.

Each type should be cylindrical with blocky ends, although rounded ends are also acceptable for slicers.

Pickling Cucumbers

"Pickling" refers to cucumbers that are primarily used for processing and pickling. Increasingly, more pickling cucumbers are being sold fresh for immediate consumption. Some consumers have a preference for the pickling type because they have thinner skins compared with slicing cucumbers. Pickling fruits are lighter green in color, shorter, thinner-skinned, and characterized by a warty surface. All commercial cultivars have either black or white spines on the fruit surface, a trait related to fruit maturity. White-spined cultivars are generally slower in their rate of development and retain their green color and firmness longer than black-spined fruits. Cultivars with black spines tend to turn yellow prematurely, especially under high temperatures, and produce larger fruits that soften with maturity. Consequently, black-spined cultivars are used for pickling in regions where summer conditions are relatively cool. White-spined hybrids have largely replaced black-spined cultivars in warmer growing regions and in areas where once-over machine harvesting is prevalent.

For processing cucumbers, the grower generally has little choice of cultivar since the processor selects and provides the cultivars to be grown. Gynoecious hybrids are grown for just about all machine harvest. These types have also replaced many of the standard monoecious types that were previously used in hand-harvesting pickling cucumbers.

- Shorter growth cycle of 50-60 days.
- high plant populations 240,000/ha (60,000/acre).
- Concentrated fruit set adapts them for once over machine harvest.
- Predominantly female types (PF).
- Some male blossoms are produced as 10-12% male pollinator seeds are mixed in with the gynoecious types or the PF types).
- An average yield is 25 t/ha (11.4 short ton or 460 bushels/acre).

Slicing (Fresh Consumption) Cucumbers

"Slicing" refers to cucumbers that are sold fresh for immediate consumption as a salad item.

Characterized by thick, uniform, dark green skins, slicing cucumbers are longer than processing types, and their thicker skins are more resistant to damage during handling and shipping.

Average yield for slicing cucumbers in e.g. North Carolina is 11-14 t/ha (200-250 bushels/acre, but better yields of 33-37 t/ha (600-650 bushels per acre) can be obtained when growing a crop on plastic which is fertigated.

Fruits for fresh market slicing are preferably long, smooth, straight, thick-skinned, with a uniform medium-dark green color. Fresh market cultivars have fewer spines than processing types. For fresh market slicers, both monoecious hybrids and gynoecious hybrids are available. Vigor, uniformity, and higher yields are some advantages of hybrids over previous open-pollinated monoecious cultivars. Regardless of how they are to be used, cultivar differences in earliness and disease resistance are also important considerations for cultivar selection.

Greenhouse Cultivars

These should have long, relatively narrow fruits, with rounded ends. Dutch greenhouse cultivars are parthenocarpic with gynoecious expression and high-yield potential, while Japanese greenhouse cucumbers are mostly monoecious. Unlike those for processing and some slicing, greenhouse types are fairly smooth-skinned.

Zucchini

Zucchini, a summer squash, produces long, slender fruit that is picked and eaten while still immature. The vines are large and require plenty of growing room, but for those who have room, zucchini is cost-effective, because one plant can produce between 3 and 9 pounds of fruit. When zucchini plants get enough sun and water, they will continue to produce fruit for several months.

Harvesting Zucchini

Harvest zucchini when it is between 4 and 8 inches long and about 1 1/2 to 3 inches in diameter. Zucchini generally takes 35 to 55 days from planting until harvest. Zucchini fruit grows rapidly -- up to 2 inches per day, so it is best to harvest them every other day during the growing season. To remove fruit, use a sharp knife to cut it from the vine. Wear gloves, if possible, because the vines have prickly stems.

Ending Harvest

The zucchini harvest will naturally end when the growing season ends, but if the plants are producing more fruit than you can use, allow a couple of fruits to mature on the vine to slow down fruit production. Another method of preventing overwhelming fruit production is to stagger seed sowing so you have a full season of growth and harvest, but are not overwhelmed with a large number of zucchini all at once.

Ways to Plant Zucchini for Optimum Growth

Plant the zucchini in an area near a water supply, so you can keep the zucchini watered regularly. Water the plants deeply, giving them 1 inch of water per week. Zucchini needs at least four to six hours of sun each day, so make sure the growing area is away from trees, buildings and other structures that could shade the plants. The plants grow well in a variety of soils, but well-drained soil is best. Avoid planting them in weedy areas.

Fruiting Problems

Poor pollination is the most common cause of poor fruit production. Zucchini have both male and female flowers on the same plant and rely on bees for pollination. If you don't have enough bees visiting your yard to successfully pollinate them, you can use an artist's paintbrush to pollinate the plant by hand or break off a male flower and rub it against a female flower. Use only freshly opened flowers and hand pollinate in the early morning. The female flowers are only receptive for one day.

Pumpkin

Pumpkins are a member of the *Cucurbitaceae* family, which also includes squash, cantaloupes, cucumbers, watermelons, and gourds. The pumpkin is undoubtedly American in origin. Fragments of stems, seeds, and fruits of *C. pepo* and *C. moschata* have been identified and recovered from the cliff dweller ruins of the southwestern United States. It is believed that *C. moschata* originated in the Mexican-Central American region and that *C. maxima* originated in northwestern South America. Cultivation of some of these pumpkins and squashes is almost as old as maize, and the presence in eastern Asia of distinct forms of squashes and pumpkins hints of distribution occurring in the sixteenth and seventeenth centuries.

Currently, production of pumpkin in the United States is more than 1 billion pounds annually, generating over $100 million in farm receipts from around 50,000 acres. Production in the northeastern United States is estimated at almost 25,000 acres.

Production Considerations

Plants can be annual or perennial vines and grow best under warm and moist conditions similar to their native semi-tropical to tropical climates. Both male and female flowers are produced on each plant and fruit shape, size, and appearance are quite variable, ranging from smooth and small (under 3 pounds) to ribbed and quite large (more than 90 pounds).

Site Selection

Pumpkins should be grown on soils that have good water infiltration rates and good water-holding capacity. If pumpkins are going to be grown on sandy soils, access to irrigation is important to obtain optimum plant growth, uniform fruit set, and development. Soil pH should be in the 5.8-6.6 range with minimum soil compaction. Pumpkins are very sensitive to cold temperatures (below 50° F) and plants and fruit will exhibit injury from even a slight frost. The best average temperature range for pumpkin production during the growing season is between 65 and 95° F; temperatures above 95° F or below 50° F slow growth and maturity of the crop. Pumpkins require a constant supply of available moisture during the growing season. Water deficiency or stress, especially during the blossom and fruit set periods, may cause blossoms and fruits to drop, resulting in reduced yields and smaller-sized fruits.

Planting and Fertilization

Pumpkins are generally seeded in the field during the first couple weeks of July. Since they are a warm-season crop, they should not be seeded until the soil temperature reaches 60° F three inches beneath the soil surface. Pumpkins seeded in cool soils may suffer from seed corn maggot injury. No-till pumpkins can be seeded with a no-till planter or transplanted in a minimally prepared bed with only secondary tillage such as an s-tine cultivator or in a previously tilled field without any tillage treatment, saving both time and labor. Because pumpkin seed germinates and develops optimally when soil temperatures are at least 60° F, early pumpkin production using no-till is difficult because of the cold soil temperatures. However, by mid- to late June, soil temperatures in a no-till field are warm enough for rapid pumpkin seed germination and growth. In addition, no-till reduces soil moisture loss early in the season and has more water available for pumpkin plant growth later in the season. If considering no-till pumpkin production, the following factors must be considered to be successful: variety, planting date, soil fertility practices, insect pressure and control, planting equipment, cover crop type and stand, and weed species and population distribution in the field.

Because pumpkins are a warm-season crop, they can also be grown as transplants on raised beds with black or silver plastic mulch and drip irrigation for optimum plant growth and yields. The use of plasticulture in the production of pumpkins will:

- Increase soil temperature 8-12 degrees warmer than bare soil

- Maintain soil water availability

- Reduce weeds

- Improve soil tilth

- Reduce fertilizer and pesticide leaching under the bed.

Use of drip irrigation also allows for fertilizer application (injection) throughout the growing season. Growing pumpkins using plasticulture will double the yield of pumpkins grown on bare soil or in no-till production.

Pumpkins are generally planted as single rows with 30-40 inches between plants in the row and 8-12 feet between rows, depending on plant type. Plant populations at these spacing are approximately 1,600 (for pumpkins in excess of 30 pounds) to 2,800 plants per acre (for pumpkins less than 8 pounds).

Fertilizer recommendations are based on soil test results, and soil tests should be taken every year. In absence of soil test results, recommended N-P-K application rates are 80-150-150 broadcast or 40-75-75 banded at planting. Soil calcium levels should be checked; if soils are testing low or low to medium in calcium and have not received any calcitic (calcium-based) lime applications, apply gypsum to the field in bands where rows will be planted prior to planting pumpkins. Gypsum will supply calcium to the soil without changing soil pH.

Pollination

Honey bees are important for proper, complete pollination and fruit set. One hive per acre is the recommended population of honey bees for maximum fruit production. Populations of pollinating insects may be adversely affected by insecticides applied to flowers or weeds in bloom.

Pest Control

Control of weeds can be achieved with a good crop rotation system and herbicides. Pumpkins can be competitive with weeds once they develop their mature canopy, if they are planted at high plant populations, or if they are planted on plastic mulch. There are several pretransplant and postemergence herbicides labeled for pumpkins, depending on specific weed problems requiring control and stage of pumpkin growth. In addition, under mild infestation levels, early cultivation (if possible prior to vine running) can minimize weed problems.

Insects can be a major problem in pumpkin production. Cucumber beetles, aphids, squash vine borer, seed corn maggot, squash bug, and spider mites have the potential to cause a reduction or loss of the marketable crop in any given year. Monitoring insect populations through scouting will help growers determine when they should start and stop spraying pumpkins and the intervals between applications.

Several diseases of pumpkin can cause a reduction in crop yields, especially bacterial wilt, viruses (powdery mildew, downy mildew), and scab. Optimum crop yields and fruit color may only be possible if a scheduled fungicide program is used to prevent leaf loss from mildews. Crop rotation, good soil and air drainage, and use of resistant varieties (where possible) can help reduce problems from these diseases in the field.

Harvest and Storage

Pumpkins are hand-harvested at their mature stage, color (orange or white), and size. Because fruit are pollinated at different times, multiple harvests over the field are quite common. Grading pumpkins for size, maturity, and pest damage before marketing is necessary to ensure a high-quality product. Maintaining pumpkin fruit in a dry, cool environment (a barn, for example) will help extend the shelf life of the crop and help maintain a non-shrunken fruit appearance.

Placing pumpkins in a well-ventilated storage area, preferably protected from rain, maintains healthy fruit for processing (pumpkin pie mix) or late sales of Jack-O-Lantern types. Pumpkins will retain good quality for approximately 2-3 months if stored at the appropriate relative humidity (50-70 percent) and temperature (50-55°F).

Calabash

Bottle gourd or calabash is a delicately flavored, Cucurbita family vegetable. It is one of the chief culinary vegetables in many tropical and temperate regions around the world.

Botanically, calabash belongs to the broader Cucurbitaceae (gourd) family of plants, in the genus: *Lagenaria*. Scientific name: Lagenaria siceraria (Molina) Standl. Some of the common names are *white-flower gourd, upo-squash (Filipino), long-squash*, etc., in the west and *doodhi* or *lauki* in the Indian subcontinent.

Bottle gourd

Bottle gourd is a fast growing, annual climber (vine) that requires adequate sunlight for flowering and fruiting. It can be grown in a wide range of soils and need trellis support for a spread.

Its intensely branched stems bear musky, deep green, broad leaves just similar as that in pumpkins, and white, monoecious flowers in the summer. After about 75 days from the plantation, young, tender, edible fruits evolve that will be ready for harvesting.

Bottle gourds come in wide range of shapes and sizes.The fruit features oval, pear-shaped or elongated and smooth skin that is light green. In the case of round or pear shaped calabash, their surface is marked by inconspicuous ridges that run lengthwise. Internally, its flesh is white, spongy and embedded with soft, tiny seeds. As the fruit begins to mature, its seeds gradually grow similar to as that in honeydew melons.

Health Benefits of Bottle Gourd

- Bottle gourd is one of the lowest calorie vegetables- carrying just 14 calories per 100 g. It is one of the vegetables recommended by the dieticians in weight-control programs.

- Fresh gourds contain small quantities of folates, contain about 6 µg/100g (Provide just 1.5% of RDA). Folate helps reduce the incidence of neural tube defects in the newborns when taken by anticipant mothers during their early months of pregnancy.

- Fresh calabash gourd is a modest source of vitamin-C (100 g of raw fruit provides 10 mg or about 17% of RDA). Vitamin-C is one of the powerful natural antioxidants that help the human body scavenge harmful free radicals, which otherwise, labeled as one of the reasons for cancer development.

- Calabash facilitates easy digestion and movement of food through the bowel until it is excreted from the body. Thus, it helps in relieving indigestion and constipation problems.

- Aslo, the vegetable is also a modest source of thiamin, niacin (vitamin B-3), pantothenic acid (vitamin B-5), pyridoxine (vitamin B-6) and minerals such as calcium, iron, zinc, potassium, manganese and magnesium.

- Bottle gourd tender leaves and tendrils are also edible and indeed contain higher concentrations of vitamins and minerals than its fruit.

Selection and Storage

Bottle gourds can be available around the season in the regions wherever suitable conditions for their growth exist. In the markets, look for fresh produce featuring tender, medium size, uniform, light green color fruit. Take a close look of its stem, which may offer a valuable hint whether the produce is fresh or aged.

Avoid those with oversize, mature, yellow-discoloration, cuts and bruise on their surface. Tiny spots on the surface, however, would not lessen their quality.

At home, store them in the refrigerator set at adequate humidity where they stay fresh for 3-4 days.

Soil and its Preparation

Bottle gourd can be grown in any types of soil. But sandy loam soils are best suited for its cultivation. The land should be prepared thoroughly by five to six ploughings.

Time of Sowing and Layout

The seed is sown from January to end of February for summer crops. June – July for rainy season crop in the plains and April in the hills. Layout is ring and basin used.

Seed Rate

The seed rate is 3 to 6 kg/ ha.

Methods of Sowing and Spacing

The seed is sown by dibbling method at spacing of 2 to 3 X 1.0 to 1.5 m. Generally three to four seeds are sown in a pit at 2.5 to 3.0 cm depth. Manures and fertilizer same as for pumpkin.

Intercultural Operation

Two to three hoeing is given to keep down the weeds during the early stage of growth. The rainy season crop is usually stalked, often trained on a bower made of bamboos and sticks.

Irrigation

The summer crop requires frequent irrigation at an interval of 4 to 5 days. The winter crop is irrigated as and when needed.

Harvesting

The fruits should be harvested when they are still green. Delay in harvesting causes the fruit to become unit for marketing.

Yield

The average yield is 90 to 120 quintal / ha.

Pod Vegetable

Pod vegetables are a type of fruit vegetables where pods are eaten, much of the time as they are still green.

Such plants as green beans or *Lotus tetragonolobus* in the family Fabaceae, or okras in the family Malvaceae are pod vegetables.

Bean

A bean is a seed of one of several genera of the flowering plant family Fabaceae, which are used for human or animal food.

Cultivation

Field beans (broad beans, *Vicia faba*), ready for harvest

Unlike the closely related pea, beans are a summer crop that need warm temperatures to grow. Maturity is typically 55–60 days from planting to harvest. As the bean pods mature, they turn yellow and dry up, and the beans inside change from green to their mature colour. As a vine, bean plants need external support, which may be provided in the form of special "bean cages" or poles. Native Americans customarily grew them along with corn and squash (the so-called Three Sisters), with the tall cornstalks acting as support for the beans.

In more recent times, the so-called "bush bean" has been developed which does not require support and has all its pods develop simultaneously (as opposed to pole beans which develop gradually). This makes the bush bean more practical for commercial production.

Types

Currently, the world genebanks hold about 40,000 bean varieties, although only a fraction are mass-produced for regular consumption.

Some bean types include:

- *Vicia*
  - ◦ *Vicia faba (broad bean or fava bean)*

Vicia faba or broad beans, known in the US as fava beans

- *Phaseolus*
  - ◦ *Phaseolus acutifolius (tepary bean)*
  - ◦ *Phaseolus coccineus (runner bean)*
  - ◦ *Phaseolus lunatus (lima bean)*
  - ◦ *Phaseolus vulgaris (common bean; includes the* pinto bean, kidney bean, black bean, Appaloosa bean *as well as* green beans, *and many others)*
  - ◦ *Phaseolus polyanthus (a.k.a. P. dumosus, recognized as a separate species in 1995)*
- *Vigna*
  - ◦ *Vigna aconitifolia (moth bean)*
  - ◦ *Vigna angularis (*adzuki bean*)*
  - ◦ *Vigna mungo (urad bean)*
  - ◦ *Vigna radiata (*mung bean*)*
  - ◦ *Vigna subterranea (Bambara bean or ground-bean)*
  - ◦ *Vigna umbellata (ricebean)*
  - ◦ *Vigna unguiculata (cowpea; also includes the* black-eyed pea, *yardlong bean and others)*

- *Cicer*
  - *Cicer arietinum* (chickpea *or garbanzo bean*)
- *Pisum*
  - *Pisum sativum* (pea)
- *Lathyrus*
  - *Lathyrus sativus (Indian pea)*
  - *Lathyrus tuberosus (tuberous pea)*
- *Lens*

Lentils

  - *Lens culinaris (lentil)*
- *Lablab*
  - *Lablab purpureus (hyacinth bean)*

Hyacinth beans

- *Glycine*
  - *Glycine max (soybean)*
- *Psophocarpus*
  - *Psophocarpus tetragonolobus (winged bean)*

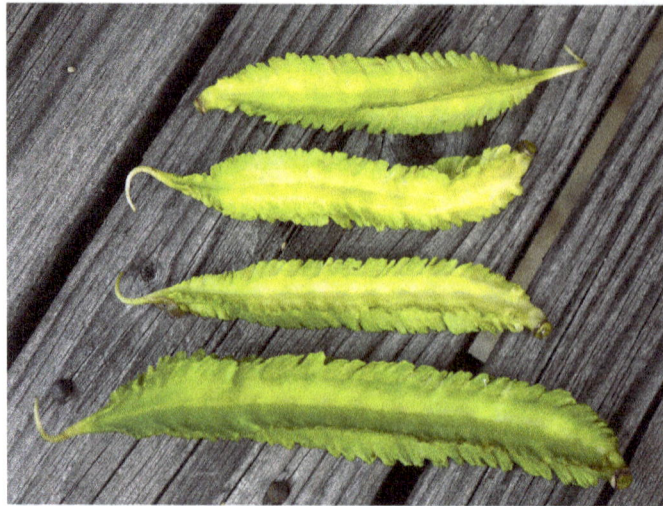

Psophocarpus tetragonolobus (winged bean)

- *Cajanus*
  - *Cajanus cajan (pigeon pea)*
- *Mucuna*
  - *Mucuna pruriens (velvet bean)*
- *Cyamopsis*
  - *Cyamopsis tetragonoloba or (guar)*
- *Canavalia*
  - *Canavalia ensiformis (jack bean)*
  - *Canavalia gladiata (sword bean)*
- *Macrotyloma*
  - *Macrotyloma uniflorum (horse gram)*
- *Lupinus* (lupin)
  - *Lupinus mutabilis (tarwi)*
  - *Lupinus albus (lupini bean)*
- *Arachis*
  - *Arachis hypogaea (peanut)*

Green Beans

Green beans come in a wide variety of sizes, shapes, and colors, and two distinctly different growing habits, so they can be grown to suit just about any garden space in most climates. And in addition to being a tasty garden treat, green beans can improve soil fertility by fixing nitrogen with their roots.

The biggest distinction that you'll need to know about before running out and buying seeds to grow your own green beans is their growth habits, which can be either pole beans (climbing vines) or bush beans (compact plants that don't need support). Pole beans are well suited to trellises, bean tipis, or along fences, as they really do need to climb up a pole of some sort, without which they sprawl on the ground and quickly become a tangled jungle that isn't conducive to optimal growing or harvesting of the beans. Bush beans, on the other hand, are much shorter plants that can stand alone without support, are often quicker to mature than pole beans, and could be grown in a container garden.

Methods to Grow Green Beans

Most green beans should be planted after the soil warms and the danger of frost is gone, and need to be planted about an inch deep (and as deep as two inches, especially in arid climates). As a rule of thumb for planting, plan for about 10 to 15 green bean plants for each person in your household. Once planted, the beds should be watered to stay evenly moist until all of the seedlings emerge from the ground, at which point the surface of the soil can be allowed to dry out between watering. Green beans will do best in fertile soil that is rich in organic matter, and digging some finished compost into the garden beds will help them thrive. Once the green bean seedlings have several true leaves, cover the garden beds with several inches of mulch to conserve moisture, keep soil temperatures cooler, and keep weed seeds from germinating.

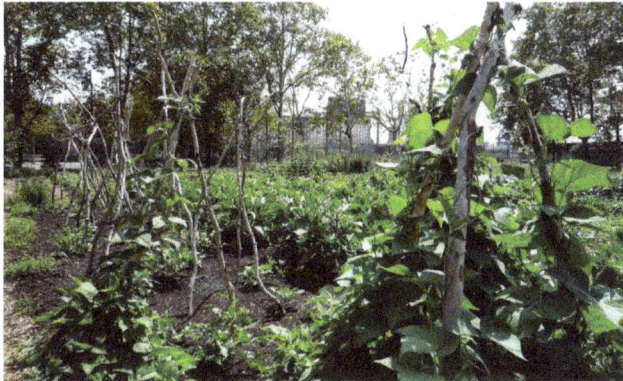

Pole Bean

Pole beans are grown commercially in the mountain counties and, on a limited scale, in a few of the eastern counties. They are produced in home gardens throughout the state. Pole beans are grown for their distinctive flavor, long pods, high yield, long harvesting season, and high price.

Soils - A well-drained, loose-textured soil is preferred. Soils that cake or crust easily result in poor stands. The soil pH should be 5.5 to 6.5, preferably in the 5.8 to 6.0 range. Well-drained bottom-lands in the mountains have been most satisfactory. To reduce soil-borne diseases, follow at least a 3-year rotation. If root rots are a problem, use a 4- to 5-year rotation. Plowing under small grain as a green manure crop will increase the organic matter of the soil as well as the yield and quality of the beans.

Irrigation

Adequate moisture (1 to 1.5 inches per week) is extremely important during pod development. Hot, dry weather after flowering begins will result in poorly-shaped pods that are tough and woody. Irrigation is the best insurance you have against dry weather damage.

Harvest

About 60 to 70 days are required from seeding to first harvest. Pole beans are usually harvested 5 times (occasionally as few as 3 or as many as 10), with about 3 to 5 days between harvests. Pole beans should be harvested before they get tough and woody; thus, timing is important.

Pea

A pea, although treated as a vegetable in cooking, is botanically a fruit; the term is most commonly used to describe the small spherical seeds or the pods of the legume *Pisum sativum*. This was the original model organism used by Gregor Mendel in his early work on genetics.

The name is also used to describe other edible seeds from the Fabaceae like the pigeon pea (*Cajanus cajan*), the chickpea, the cowpea (*Vigna unguiculata*)and the seeds from several species of *Lathyrus*.

P. sativum is an annual plant. It is a cool season crop, planted in winter. The average pea weighs between 0.1 and 0.36 grams. The species is as a fresh vegetable, but is also grown to produce dry peas like the split pea. These varieties are typically called field peas.

P. sativum has been cultivated for thousands of years, the sites of cultivation have been described in southern Syria and southeastern Turkey, and some argue that the cultivation of peas with wheat and barley seems to be associated with the spread of Neolithic agriculture into Europe.

Selection of Land

Pea crop can be grown on a wide range of soils but it grows well on well- drained loamy soil. Water logging is injurious to the crop. If the soil is acidic, liming has to be done after soil analysis.

Preparation of Land

Thoroughly prepared seed bed is required for pea cultivation. First ploughing should be given by soil inverting ploughs like mould board plough. Two subsequent ploughings accompanied by planking will be sufficient to obtain desired tilth of the seed bed. Care should be taken to provide gentle gradient so that excess water may not stand in the field.

Sowing Time

Seed can be sown from 15th October to ending November.

Seed Rate

The seed rate of 75 to 80 kg /ha is recommended. In case of bold seeded varieties, seed rate can be increased up to 100 kg/ha.

Sowing of Seed

Line sowing with the help of seed drill or opening the furrows at 30-40 cm between the rows. The seed should be placed 5 to 6 cm deep in the soil. Seeds may be inoculated with *Rhizobium* culture for quick fixation of atmospheric nitrogen.

Fertilization

Application of 20 kg/ha of nitrogen as a starter dose will be sufficient to meet the nitrogen requirement of the crop in the initial stage. It will be advisable to apply the fertilizers as per the results of soil analysis. However, in absence of soil test, it would be advisable to apply 50 kg each of P_2O_5 and K_2O per ha as basal dose.

Irrigation

During the growth period, peas should receive at least two irrigations during the month of May/June, if rainfall does not occur.

Harvesting and threshing

At maturity, all the leaves turn yellow and fall down leaving behind stalks with pods. Care should be taken while threshing is done i.e, not to over beat or over trample which may damage the seed coat and reduce quality. Moisture content of 13 to 14 % in seed is ideal for threshing purpose.

Storage

The grains should carefully be dried before storage to ensure that moisture content does not exceed 10%. The seed can be stored in dry bins or in bags kept on wooden racks under cool and dry conditions.

Common Diseases of Peas and their Control

Name of the host	Disease and casual organism	Symptoms	Control measures
Pea	Powdery mildew (*Erysiphy polygoni*)	The first signs are in the form of faint, slightly discolored, tiny specks which spread to form variously sized areas and over run the leaf,stem or pod. Underneath this, the pod or leaf may assume a brown or purplish color and in the powdery area dark to black cliestothecia (pin heads) may form.	Spray the crop thrice with Sulfur fungicide such as Sultaf or Suprsol at 0.25% concentration at an interval of 15 days starting after the appearance of disease.

Common Inset Pests of Peas and their Management

Name of the pest	Nature of damage	Period of activity	Management
Pod borer	Young larvae feed on tender portion of leaves and shoots. Grown-up larvae bore into pods and feed on developing grains.	April to July	For peas, spray Endosulfan 0.05 to 0.07 % or Carbaryl 0.15 to 0.2% at the onset of first flush of flowering and repeat after 25 days interval, if necessary.

Leaf miner	The larvae mine leaves making prominent whitish tunnels resulting in the loss of green matter and also affecting flowering and pod formation.	March to June	Soil application of Carbofuran 3G 32 kg/ ha will take care of aphids, white grubs and leaf miners.
Aphids	Suck sap from tender shoots of the plants resulting in stunted growth. Leaves may sometimes turn yellow and dry up. The formation of flowers and pods is adversely affected.	April to July	In case peas are to be consumed as fresh vegetable, apply Dichlorvos 0.05 %.
White grubs	Grubs damage the roots	March to July	a) Proper tillage operations before sowing will expose the grubs and other soil insects to their natural enemies and weather conditions. b) Treat the soil with carbofuron3G granules 25-30 kg/ha or Carbaryl 10% dust 25-30 kg/ha before sowing of seed in case of heavy pest infestation.

Snow Pea

Snow peas are also known as Chinese pea pods since they are often used in stir-fries. They are flat with very small peas inside; the whole pod is edible, although the tough "strings" along the edges are usually removed before eating. Snow peas are mildly flavored and can be served raw or cooked.

Snow peas require adequate air circulation to the beneficial nitrogen-fixing bacteria that live on the plant root. Constant air supply is ensured by avoiding waterlogged areas and not compressing the soils after planting

Snow peas prefer a fertile Sandy loam that drains well, but tolerate most soils except impermeable clay. A Ph level of 6.0-7.5 is preferred. In case of soils with lower Ph, 1 kg of humipower can be mixed with 50kg fertilizer or manure. This raises the soil Ph since low soil Ph has adverse effects on crops such as;

- Toxicity of Aluminum which becomes soluble.

- Affects nutrient availability.

- Leaching of Mg, K, and Ca since they become soluble.

- Lack of nodulation of legumes.

They thrive well in cool weather in upper and lower highlands at altitude of between 1500-2600 m above the sea level and temperatures of between 12°C – 20°C with well distributed rainfall of 1500 mm-2100 mm per year.

Pests

Aphids(Acyrthosiphonpisum)

Aphids are very common sap-sucking insects that can cause a lack of plant vigor, distorted growth and often secrete a sticky substance called honeydew which allows the growth of sooty moulds.

This Sooty moulds coats the leaves surface blocking stomata which acts as entry for carbon dioxide for photosynthesis hence lack of the crucial process in the plant leading to stunted growth. Aphids also vector many diseases even in other crops.

Aphids are controlled through spraying Kingcode Elite 10 ml/20 l.

Thrips(Tabacispp)

Thrips are tiny slender insects with fringed wings. They are sucking pests feeding by puncturing epidermal layers of host tissue resulting in silvering of leaves' surface.

Control is by use of insecticide; Alonze 3 ml/20 l.

Pea Weevil (*Bruchuspisorum*)

These are small, black to brownish insects with a white zigzag running across the back.

The pea larvae hatch and burrow into the pods and feed on the developing peas.

They can be controlled by use of insecticide; Pentagon 10 ml/20 l.

Cutworms (Agrarian *Segetum*)

Cutworms are more larvae that hind under the litter or soil during the day, coming out at the dark to feed on the plant. The larvae attacks the crop at the stem base by cutting it down hence the name 'cutworm'.

Control is done by drenching with Pentagon 20 ml/20 l.

Diseases

Damping Off and Root Rot

It is caused by a number of pathogens such as *pythium* and *rhizoctonia*. It is exacerbated by cool wet soils. Seeds becomes soft and rolled while seedling fall due to sunken lesion. Older seedlings develop root rot when peas are planted in overly wet soils.

Symptoms:

Wilting, Foliage becomes brown, Roots becomes brown, Stunted growth

Seedling attached by damping off

Management:

In addition to use of certified seeds a drench of Pyramid 700 WP 100 g/20 l is used to control damping off.

Downy Mildew

It is caused by *Peronosporaviciae*. The disease survives in soils and on plant debris. It can also be seed borne. The disease develops quickly in cold conditions (5°-15°C) and wet for 4 -5 days. This often happens when seedlings are in early vegetative stages. Rainfall is the major method of spores dispersal and secondary infection.

Symptoms:

The underside of leaves are covered with a fluffy mouse grey spore mass.

Sickly yellowing green appearance.

Deformation of pods covered with yellow and brownish areas.

Downy mildew in snow peas.

Management:

In addition to use of certified seeds, a fungicide, Gearlock turbo 25 g/20 l can be sprayed to control the disease.

Powdery Mildew

It is caused by *Erysiphepisi*. Unlike downy mildew, it is prevalent in days of warm weather.

Symptoms:

Covering of infected Plants with white powdery film.

Severely affected leaves turns blue-white in color.

Powdery mildew in snow peas leaves.

In addition to seed treatment and crop rotation, fungicides such as Chariot 20ml/ 20l or Ransom 10 g/20 l is sprayed to the affected crop to eradicate the disease.

Nutrition

Nutrition or fertilizer application is determined by soil analysis. However, up to 10 tons of farm yard manure should be applied. Applications of DAP fertilizer at a rate of 250 kg per ha at sowing time and again after one month is recommended for root growth. At flowering stage the plants should be dressed with CAN at a rate of 200 kg per ha. All fertilizer applied should be mixed well with soil.

Avoid excess nitrogen which will promote vegetative growth at the expense of growth of pods. Hand weeding is recommended since the crop has shallow roots and care must be taken not to injure the roots. Alternatively, after using DAP in sowing gatit range can be used in subsequent fertilizer spraying.

Harvesting

Harvesting of snow peas is determined by horticultural harvesting index rather than maturity index. They are harvested when pods start to fatten, but before peas get too large. For best flavor, cook or freeze peas within a few hours of picking.

Okra

Okra, (Abelmoschus esculentus) is a herbaceous hairy annual plant of the mallow family (Malvaceae). It is native to the tropics of the Eastern Hemisphere and is widely cultivated or naturalized in the tropics and subtropics of the Western Hemisphere for its edible fruit. The leaves are heart-shaped and three- to five-lobed. The flowers are yellow with a crimson centre. The fruit or pod, hairy at the base, is a tapering 10-angled capsule, 10–25 cm (4–10 inches) in length (except in the dwarf varieties), that contains numerous oval dark-coloured seeds. Only the tender unripe fruit is eaten. It may be prepared like asparagus, sauteed, or pickled, and it is also an ingredient in various stews and in the gumbos of the southern United States; the large amount of mucilage (gelatinous substance) it contains makes it useful as a thickener for broths and soups. The fruit is grown on a large scale in the vicinity of Istanbul. In some countries the seeds are used as a substitute for coffee. The leaves and immature fruit long have been popular in the East for use in poultices to relieve pain.

Soil Nutrition

Okra is slightly sensitive to soil acidity but generally soil pHs of 6.0 to 7.0 are satisfactory. Fertilization rates vary dramatically with soil type and should be calculated based on soil test results. Okra is considered a heavy feeder and will and remove approximately 180 lbs/acre N, 30 lbs/acre phosphate, and 150 lbs/acre of K per season. Okra generally responds well to fertilization when soil fertility is low. Side dressings of N may be applied 2 weeks before harvest and again after harvest has begun to stimulate additional vegetative growth.

Isolation

Recommended isolation distances vary from one country to another, but a minimum isolation distance of 500 m is desirable.

Planting

In the southern US, okra is generally direct seeded when soil temperatures exceed 60 degrees F. The seed is planted thickly in rows spaced 2.5 to 4.0 feet part. When plants are established they are thinned to an in-row spacing of 12 inches apart for dwarf cultivars and 18-24 inches for larger cultivars. Okra seeds have hard seed coats an soaking seeds in hot water for 1-2 hours prior to planting improves germination. In northern areas okra is sometimes transplanted but this is not an established practice for large scale commercial production. Poor stand establishment can be a problem when soil temperatures are cold or because of the hard seed coats.

Irrigation

Okra plants have good drought tolerance but yield is adversely affected when plants are water stressed during flowering. During flowering and fruit development, uniform soil moisture is essential. During flowering and fruit set plants should receive at least 1.5 inch per week of water to ensure fruit set and development. In western states, furrow irrigation is used. In eastern states drip irrigation is gaining popularity because it helps reduce foliage diseases and overhead irrigation are used as needed.

Roguing

1. *Before flowering:*

    Check the general plant height and habit, pigmentation of leaves, petioles and stems; remove plants with virus symptoms.

2. *Flowering* check:

    The relative size and color intensity of flowers; remove plants with virus symptoms.

3. *Fruiting* check:

    That fruit is true to type; remove plants with virus symptoms:

Diseases	Insect Pests
Since okra is related to cotton, many of the pests are the same. Okra should not be rotated with cotton. Cotton diseases include: • Root rot • Pod rot • Fusarium wilt • Verticillium wilt • Powdery mildew • Dry rot • Cercospera blight and leaf spot • Alternaria leaf spot	• Boll weevil • Pink boll worm • Corn earworm • Flea beetle • Cotton aphid • Stemborer • Leafhopper • Spidermite • Whitefly

Seed Harvest

There is a sequential ripening of okra pods on a plant. The fruits of the angular fruited types have a tendency to split when the seed ripens. The maturity of seeds is associated with the pods becoming gray or brown according to the cultivar.

The traditional hand harvesting of ripe pods is still done in many tropical areas where there is adequate labor, although the crop is combined in the USA.

Cleaning

Seeds are extracted after the hand-harvested pods become dry and brittle. The most efficient method of hand seed-extraction is to twist the pods open. Alternatively the pods are either flailed or the seeds are extracted with a stationary thresher.

In some areas of the world, especially Malaysia, the initial pods are harvested as a fresh vegetable and the later pods are retained on the plants for seed production. This practice probably does not reduce the potential seed yield in areas with a sufficiently long growing season, as removal of early pods tends to encourage further extension growth and flower development.

Seed Yield

A relatively high seed yield of 1500 kg/ha (1,338 pounds per acre) is achieved in the okra seed producing areas of the USA but in many tropical countries the yield rarely exceeds 500 kg/ha (446 pounds per acre).

References

- Daniel Zohary and Maria Hopf Domestication of Plants in the Old World Oxford University Press, 2012, ISBN 0199549060, p. 114.

- Guide-capsicum-production-technologies, agripedia: krishijagran.com, Retrieved 19 June 2018

- Harrison, DC; Mellanby, E (October 1939). "Phytic acid and the rickets-producing action of cereals". Biochem. J. 33: 1660–1680.1. doi:10.1042/bj0331660. PMC 1264631. PMID 16747083

- Crop-guide-cucumber, cucumber-fertilizer: haifa-group.com, Retrieved 31 March 2018

- "Foodborne Pathogenic Microorganisms and Natural Toxins Handbook: Phytohaemagglutinin". Bad Bug Book. United States Food and Drug Administration. Archived from the original on 9 July 2009. Retrieved 11 July 2009

- Long-zucchini-plant-produce-55042: homeguides.sfgate.com, Retrieved 12 March 2018

- Chazan, Michael (2008). World Prehistory and Archaeology: Pathways through Time. Pearson Education, Inc. ISBN 0-205-40621-1

- How-grow-green-beans, lawn-garden: treehugger.com, Retrieved 22 May 2018

- Gorman, CF (1969). "Hoabinhian: A pebble-tool complex with early plant associations in southeast Asia". Science. 163 (3868): 671–3. doi:10.1126/science.163.3868.671. PMID 17742735

- Whats-the-difference-between-snow-peas-sugar-snap-peas-and-english-peas-ingredient-intelligence-205118: thekitchn.com, Retrieved 14 June 2018

- Vicky Jones (15 September 2008). "Beware of the beans: How beans can be a surprising source of food poisoning". The Independent. Retrieved 23 January 2016

- Snow-pea-production: farmlinkkenya.com, Retrieved 25 July 2018

Stem Vegetables and Inflorescence Vegetables

Stem vegetables are stems of plants that are eaten as vegetables. Plant inflorescences, such as flowers and flower buds and their stems and leaves that are eaten as vegetables are called inflorescence vegetables. This chapter closely examines the key concepts of stem vegetables and inflorescence vegetables, and their various types.

Stem Vegetables

Stem vegetables (vegetable crops) are those plants from which edible botanical stems are harvested for use in culinary preparations. They can be divided further into those with edible stems that are above ground and those with modified underground stems. Swollen modified stems, such as bulbs, tubers, corms, and rhizomes, serve as main food storage organs.

Crop plants which are grown for their starchy roots, tubers and corms are called root and tuber crops which are generally placed under the domain of agronomy. But as to human consumption, they can be considered vegetables under horticulture.

Examples of these vegetable crops, without considering geographical adaptation, are provided in the table below. The botanical names, family, and other relevant information are supplied. The names of the underground storage organs of root and tuber crops were checked with the list provided by Kawakami and the current family names were confirmed with Simpson.

Table: List of stem vegetables with their scientific names, family, and other information.

Crop name	Scientific name	Family	Collective name for members of the family, other info
Stem Vegetables			
Examples of stem vegetables with edible aboveground stems:			
Asparagus	*Asparagus officinalis*	Asparagaceae	Asparagus family, but formerly under Liliaceae (Merrill 1912); the edible part is the young shoot commonly called "spear," best consumed when the tip is still tightly closed.
Bamboos	various species	Poaceae/ Gramineae	Grass family; the edible part is the young, newly emerged shoot.
Kohlrabi	*Brassica oleracea var. gongylodes*	Brassicaceae/ Cruciferae	Mustard family, also called Cole Crops and Crucifers; the main consummable plant part is the basal stem which forms a spherical structure.

Potato vine, kangkong	*Ipomoea aquatica*	Convolvulaceae	Morning Glory/Bindweed family; both stems and leaves are eaten cooked or blanched.
With edible modified underground stem called bulb:			
Chive	*Allium schoenoprasum*	Amaryllidaceae	Amaryllis family; formely under Liliaceae (Lily family, Merrill 1912) but Simpson (2010) preferred it under Alliaceae (Onion family or alliaceous crops); Alliaceae has been placed within an expanded Amaryllidaceae by the Angiosperm Phylogeny Group III (APG III, 2009).
Garlic	*Allium sativum*	Amaryllidaceae	Amaryllis family
	Allium porrum	Amaryllidaceae	Amaryllis family
Onion	*Allium cepa*	Amaryllidaceae	Amaryllis family
Shallot	*Allium cepa,* Aggregatum group	Amaryllidaceae	Amaryllis family
With edible modified underground stem called tuber:			
Jerusalem artichoke	*Helianthus tuberosus*	Asteraceae/ Compositae	Sunflower or Aster family
Potato	*Solanum tuberosum*	Solanaceae	Nightshade family, also called Solanaceous crops
Yam, ube	*Dioscorea alata*	Dioscoreaceae	Yam family
Asiatic yam, tugui, apali, tam-is	*Dioscorea hispida*	Dioscoreaceae	Yam family
With edible modified underground stem called corm:			
Taro, gabi	*Colocasia esculenta*	Araceae	Arum family; some varieties are grown for their edible leaves and petioles and modified stems (stolons)
Yautia, tannia, bisol, karlang, palauan	*Xanthosoma sagittifolium*	Araceae	Arum family

Common Healthy Stem Vegetable

Asparagus

Rich in proteins and vitamins like K and C, asparagus is one of the healthiest stem vegetables you can chew on. It aids in weight loss along with regulating your blood sugar.

Bamboo Shoot

Did you know bamboo stem or the shoot is the best to treat cancer. The properties present in the shoot like lignans help to fight cancer cells.

Garlic

Garlic contains a compound called allicin which is present in it's stem. When you consume garlic you are providing your body with properties that will aid in reducing high blood pressure along with weight loss.

Onion

Onion stem is healthy as it helps to reduce inflammation in the body. It also prevents ailments and when consumed this healthy vegetable aids to promote in weight loss.

Potato

Though many refrain from this vegetable due to it's calorie rate, it is said that the potato stem is good for health. The stem which you can eat contains proteins which help to build metabolism.

Ginger

The root of the plant is beneficial for health and so is the stem. Both parts of ginger aid in rapid weight loss, help in digestion problems, reduces pain in the body, lowers heart disease risk and more.

Broccoli

Broccoli is the world's healthiest foods. Consuming the stem of this vegetable you are providing your body with a whole lot of nutrients along with proteins, that aid in regulating blood pressure, diabetes and preventing cancer too.

Cauliflower

It is a very good source of choline, dietary fiber, omega-3 fatty acids. All of these are mainly present in the stem of the vegetable. Consuming this green yummy veggie, you will be at less risk of cancer.

Celery

It regulates the body's alkaline balance, thus protecting you from problems such as acidity. Chew on fresh celery stems to avoid any acidity problem.

Leeks

Leeks help to prevent rheumatoid arthritis and type II diabetes. Consuming this high vitamin K stem food you will also increase your immunity levels.

Asparagus

Asparagus is a genus of the family Asparagaceae (formerly in Liliaceae) with more than 200 species native from Siberia to southern Africa. Best known is the garden asparagus (*Asparagus officinalis*), cultivated as a vegetable for its succulent spring stalks. Several African species are grown as ornamental plants.

Asparagus species may be erect or climbing, and most of the species are more or less woody. The rhizomelike, or sometimes tuberous, roots give rise to conspicuous fernlike branchlets. True leaves are reduced to small scales. Many species are dioecious (individuals are either male or female), and the small greenish yellow flowers in the spring are followed by red berries in the fall. Members of the genus are characterized by the presence of cladodes, which are leaflike organs in the axils of the true leaves.

Garden asparagus, the most economically important species of the genus, is cultivated in most temperate and subtropical parts of the world. As a vegetable, it has been prized by epicures since Roman times. It is most commonly served cooked, either hot or in salad; the classic accompaniment is hollandaise sauce. In 2011 the world's leading producers of asparagus were China, Peru, Germany, Mexico, and Thailand. Commercial plantations are not undertaken in regions where the plant continues to grow throughout the year, for the shoots become more spindly and less vigorous each year; a rest period is required. Where the climate is favourable and with proper care, an asparagus plantation may be productive for 10 to 15 years or longer. The best soil types for asparagus are deep, loose, light clays, with much organic matter, and light, sandy loams. Asparagus will thrive in soils too salty for other crops, but acidic soils are to be avoided. The asparagus cutting season varies from 2 to 12 weeks, depending on age of the plantation and on climate.

In parts of France, most notably at Argenteuil, asparagus is customarily grown underground to inhibit development of chlorophyll. This white asparagus is prized for its tenderness and delicate flavour. In classic French culinary nomenclature, the word "Argenteuil" denotes an asparagus garnish.

Some poisonous species are prized for their delicate and graceful foliage. *A. plumosus,* tree fern, or florists' fern (not a true fern), has feathery sprays of branchlets often used in corsages and in other plant arrangements. *A. aethiopicus* (Sprenger's fern), *A. asparagoides* (bridal creeper), and *A. densiflorus* (asparagus fern) are grown for their attractive lacy foliage and are common ornamentals.

Several species of *Asparagus* are threatened in their natural habitats. Habitat fragmentation in the Canary Islands has lead to the listing of two species (*A. fallax* and *A. nesiotes*) as endangered and two (*A. arborescens* and *A. plocamoides*) as vulnerable. *A. usambarensis* of Tanzania is also listed as endangered.

Asparagus Production

Generally, spears are harvested when they are 7 or 9 inches in length and green in color. Varieties with purple spears have also been developed by plant breeders. Excluding light when spears are emerging will produce blanched, or white, spears.

Asparagus is believed to be indigenous to parts of Russia, the Mediterranean region, and the British Isles. It was first cultivated by the early Romans, who used the asparagus for food and medicinal purposes. It was cultivated in England over two thousand years ago and brought to America by the early colonists. However, asparagus was not extensively planted by commercial growers until after 1850.

Most of the asparagus harvested in the United States is sold as fresh produce. The United States produces around 25,000-30,000 acres of asparagus with a value of $80-100 million. U.S. acreage is currently only about one-third of what it was 15 years ago due to increased imports from Central and South America.

Site Selection

Asparagus should be grown on well-drained soils that have good water-infiltration rates and good moisture-holding capacity. The soil should not be compacted and the pH should be 6.2 to 7.0. Growers should avoid planting asparagus in fields where it has been grown in previous years. Asparagus is an allelopathic species--it produces and releases toxic chemicals that inhibit and suppress the growth of young asparagus transplants or crowns. In addition, asparagus is extremely susceptible to Fusarium root rot, a soil fungus that will weaken the plant. Fusarium can survive up to seven years in infected soil and soil fumigation is not effective in reducing long-term Fusarium populations in the soil. Asparagus is extremely salt tolerant.

Planting and Fertilization

Commercially, asparagus can be started in the greenhouse 8 to 10 weeks prior to transplanting in the field or planted as one- or two-year-old crowns. Crowns are developed root systems with a fairly defined storage organ and growth buds.

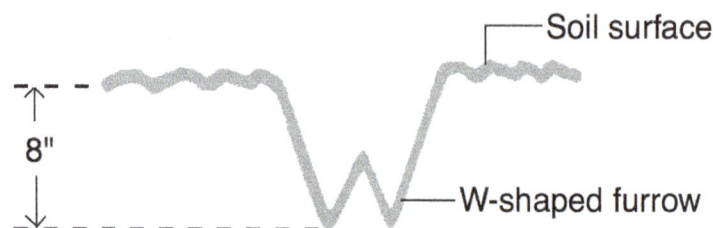

Growers generally plant approximately 12,000 to 14,000 plants per acre in single rows, with 12 inches between plants in the row and 5 to 6 feet between rows. Whether planting crowns or transplants, the asparagus is planted in an 8-inch-deep furrow with a W-shaped configuration at the bottom of the furrow. The crown and transplant are planted in the W-shaped furrow beneath the soil surface, and the furrow is gradually filled with soil during the growing season. Asparagus usually is planted in May so that extensive foliage (fern) develops before winter.

Fertilizer recommendations should be based on annual soil test results. In absence of soil test results, the recommended N-P-K application rates are 50-100-150 pounds per acre broadcast in the spring of every year before spear emergence.

Pest Control

Weed control can be achieved with a good crop rotation system, herbicides, and straw mulch. Several preplant and postemergence herbicides are available for asparagus, depending on the specific weed problem and the time of year. If infestation levels are light, early cultivation (prior to spear emergence) can help reduce weed problems.

Insects can be a major problem in asparagus production. Asparagus beetles, asparagus aphids, cutworms, and Japanese beetles all can cause crop losses. Monitoring insect populations will help you determine when you should use pesticides and how often you should spray.

Several asparagus diseases can reduce crop yields, especially Fusarium root rot and rust. These diseases can be prevented by having a good crop rotation system, planting in soil with good water and air drainage, and using disease-resistant varieties.

Irrigation

Irrigation is highly recommended and will help ensure a more consistent crop from year to year. Trickle irrigation is greatly preferred over overhead irrigation because it adds water directly to the root zone and does not wet the fruit. Also, very little water is lost from evaporation.

Harvest and Storage

The length of time for harvesting asparagus each year increases gradually until the plants reach full maturity (5 years). The first year after planting, asparagus can be harvested for about 7 days; the second-year harvest period lasts for about 14 days; the third-year harvest period is about 3 weeks; the fourth-year harvest period is 30 to 36 days; and by year five (when the plants have reached full maturity) the harvest period is approximately 6 to 7 weeks. Asparagus spears can be cut with an asparagus knife or snapped off near the soil line. Spears are harvested when they reach at least 7 inches in height and have a spear diameter of at least 5-16 inch. When growing under seasonal temperatures, asparagus should be harvested every day since spears can increase in length by as much as 2 inches per day. Harvesting asparagus when it is greater than 12 inches in length (spear diameter becomes thinner) will reduce the total marketable harvest over the life of the planting.

Proper postharvest handling of asparagus is critical to marketing success. You should cool the picked asparagus immediately after harvest to remove field heat and improve shelf life. Refrigeration immediately after harvest will help guarantee high quality. Asparagus that is maintained at 32 to 36° F and 90 to 95 percent relative humidity will retain good quality for approximately 7 to 14 days.

Health Benefits of Asparagus

- Asparagus is a very low-calorie vegetable. 100 g fresh spears carry just 20 calories.

- Besides, its spears contain moderate levels of dietary fiber. 100 g of fresh spears provide 2.1 g of roughage. Dietary fiber helps control constipation conditions, decrease bad (LDL) cholesterol levels by binding to it in the intestines and regulate blood sugar levels. Studies suggest that high-fiber diet help cut down colon-rectal cancer risks by preventing toxic compounds in the food from absorption.

- Its shoots have long been used in many traditional medicines to treat conditions like *dropsy* and *irritable bowel syndrome.*

- Fresh asparagus spears are a good source of anti-oxidants such as *lutein, zeaxanthin, carotenes,* and *cryptoxanthins.* Together, these flavonoid compounds help remove harmful oxidant free radicals from the body protect it from possible cancer, neurodegenerative diseases, and viral infections. Their total antioxidant strength, measured regarding oxygen radical absorbance capacity (ORAC value), is 2150 μmol TE/100 g.

- Fresh asparagus is rich sources of folates. 100 g of spears provide about 54 μg or 14% of RDA of folic acid. Folates are one of the essential co-factors for the DNA synthesis inside the cell. Scientific studies have shown that adequate consumption of folates in the diet during pre-conception period and early pregnancy helps prevent neural tube defects in the newborn baby.

- Its shoots are also rich in the B-complex group of vitamins such as thiamin, riboflavin, niacin, vitamin B-6 (pyridoxine), and pantothenic acid. These groups of vitamins is essential for optimum cellular enzymatic and metabolic functions.

- Fresh asparagus also contains fair amounts of antioxidant vitamins such as vitamin-C, vitamin-A, and vitamin-E. Regular consumption of foods rich in these vitamins helps the body develop resistance against infectious agents and scavenge harmful, pro-inflammatory free radicals from the body.

- Its shoots are also an excellent source of vitamin-K. 100 grams carry about 35% of DRI. Vitamin-K has potential role bone health by promoting bone formation activity. Adequate vitamin-K levels in the diet help limiting neuronal damage in the brain; thus, has established a role in the treatment of patients with Alzheimer's disease.

- Asparagus is an excellent source of minerals, especially copper and iron. Also, it has small amounts of some other essential minerals and electrolytes such as calcium, potassium, manganese, and phosphorus. Potassium is an important component of cell and body fluids that helps controlling heart rate and blood pressure by countering effects of sodium. Manganese used by the body as a co-factor for the antioxidant enzyme, *superoxide dismutase.* Copper required in the production of red blood cells. Iron is essential for cellular respiration and red blood cell formation.

Celtuce

Also known as stem lettuce, asparagus lettuce or celery lettuce, celtuce (*Lactuca sativa* var. *augustana*) is a specific variety of lettuce grown for its thick, fleshy stems. A popular Asian vegetable, celtuce has yet to be adopted by many North American cooks, but once they give it a go, they won't look back.

While the leaves can be eaten in the spring, just like any other type of lettuce, celtuce is prized for its tender white stems. Eaten cooked or raw, celtuce can be prepared in many different ways. Its mild, nutty flavor is delicious grilled or stir-fried. The stems can also be roasted or even pickled.

Growing Celtuce

In northern areas, celtuce seeds are best sown directly into the garden in mid-spring. In the South, seeds should be planted in the autumn for winter harvests. Once the seedlings reach a few inches tall, thin them to 10 to 12 inches apart. Choose a site with full sun exposure and well-drained soil, rich in organic matter.

Although the leaves can be harvested at any time for use as a fresh green, for the best stem production, allow the plants to grow throughout the summer. The thick stems are then harvested in mid to late summer, typically in early August here in my Pennsylvania garden.

Harvesting Celtuce

At harvest time, celtuce stems should be about 1 inch in diameter and 12 to 18 inches tall. It is observed that picking celtuce in the morning results in a crisper texture than afternoon harvests.

To harvest celtuce stems, snap or cut them off at ground level, leaving the uppermost leaves intact but tugging off the lowest leaves. By mid-summer, these lower leaves are often a bit bitter, but unlike the leaves, the stems of celtuce do not grow bitter in summer's heat. The flavor remains mild and delicious.

To prepare celtuce, peel the stem with a knife to reveal the inner white flesh. It can be sliced and served raw in salads and slaws, and even after cooking; it remains crispy, with a texture much like kohlrabi.

At the end of the growing season, should your celtuce go to flower, let it do so. It will drop seeds that will grow into next year's crop without any help from you.

Bamboo Shoot

Bamboo is a very economical crop. Many Chinese farmers grow it; they now it is used to make lots of useful products, and that is used frequently in China's culinary. One advantage of growing these tall thin stalks is that they do not break easily. They have high tensile strength and can bend under heavy loads; and they bounce back when the load is removed.

The Chinese admire bamboo. Sometimes, when luck is down, with much work one can eventually overcome hardship. Surviving and prospering is akin to the bamboo plant. This is why it is a popular topic for Chinese poets. Since it is a very pretty plant, it is also a favorite subject of many Chinese artists. Because this plant tastes good and has an interesting flavor and fine texture, it has many uses by those whose art is in the kitchen.

Bamboo shoots and their tall reeds require little care. They grow on marginal land such as hillsides and in places where few other crops grow. Groves of bamboo make effective windbreaks, especially for protection against winter's north and west winds. This is a native of Southern and Western China, and it is a highly profitable product.

The Chinese claim that bamboo has more than one thousand uses. They sometimes call it the the 'universal provider.' It is used to make furniture, chopsticks, sleeping mats, scaffolding, boat masts, fencing material, baskets, cooking utensils, pens for writing, farmer's hats, combs, fishing poles, garden rakes, musical instruments, blinds, fans, laminated dishes, ropes, rafts, floats, traps, snares, roofing tiles, and more. And for cooking, its leaves can be used to wrap food for steaming, and last but not least, its shoots are a gourmet delight and a favorite food.

Bamboo is fast growing and can replenish itself in a very short time. Sometimes it can grow four feet in a day. That is true of many types of bamboo including those that are small and grow no more than ten to fifteen feet tall, their main stems no bigger than a thumb. Others grow to twenty or more than thirty feet tall, the cross sections of their main stems can be three to four inches thick.

When a clump of bamboo grows, the roots form a strong web that spread and cover a large area. In the spring, after March and April rains, buds or shoots from the base of older canes or from an underground stem or rhizome push up above the ground. These shoots can be broken off and collected; they are called *hun shun* or 'spring shoots.' There are bamboo shoots that appear in the winter. They are slightly smaller and have finer textured shoots. They possess more flavor than do the spring ones and they are highly prized. These are called winter shoots or *ung shun*. They are not as plentiful as the spring shoots, but they are more in demand, and as such command a higher price. The most common and readily available ones are the sprng shoots. When in season, harvesters can hardly keep up with their preservation schedule.

There are more than a hundred types of bamboo but only ten produce shoots that are considered edible. To keep any bamboo plant from spreading, dig a shallow trench around them. This appears to stop their natural spread.

The shoots from larger bamboo types can be as long as ten inches and four to five inches thick. Those from the Zhejiang Province are deemed to have the best quality. Those from in and around Ningbo are from a smaller bamboo plant and their shoots are known as *ien shun* or 'whip shoots,' also as *ian shun* or 'arrow shoots.' These are about the size of asparagus. The Ningpo people like to preserve theirs in oil. The largest bamboo groves are mainly south of the Yangzi River in the provinces of Zhejiang, Jiangsu, Sichuan, Jiangxi, Guangdong, Guangxi, Yunnan, Hunan, and Hupeh.

In order to prepare bamboo shoots for eating, the outer husks are removed exposing the tender inner core. This core is edible. However, fresh bamboo shoots often contain a toxin, hydrocyanic acid, which can be easily removed by par-boiling them. When purchased, they usually are boiled. When prepared in a dish, bamboo gives a delicious savory taste and a refreshing crunchiness. Vegetarians highly treasure bamboo shoots because of their flavor and their texture. Lots of non-vegetarians do, as well.

Many bamboo shoots are produced in the hinterlands of China, their harvesting time a short two to three week period. If they are not picked and processed during that period, they turn into woody bamboo and are no longer edible. Harvesters often do not have appropriate transportation facilities to ship them to market immediately after they cut them. Since they can not control nor slow down the arrival of the spring shoots, they must preserve them through salting, drying, canning, pickling, freezing, etc.

In America, the products used are mostly those canned in water. These canned products retain most of the shoot's crunchiness, but they are bland. During the canning process they lose most of their original savory fresh flavor. There are other ways they can be preserved, but these are not as commonly available in the stores and most Americans would not know how to use them.

Because of the many different types of bamboo in China, variation in shoots and by-products are many. For instance, *bien jian* or 'flat sharp shoots,' as they are known, are salted and pressed flat. *Mao shun* or 'hairy shoots' come from a plant known as the 'hairy bamboo.' It is mainly grown on mountain slopes. These are the largest of all bamboo species, their joints and segments huge and probably the strongest of any of the bamboo varieties. They are used a lot for scaffolding, their shoots hairy and thus their name. These shoots are not as tasty nor is their texture as fine as shoots of smaller types.

Bamboo shoots grown in the province of Guangdong are known as *ling nan*; they have a slightly bitter taste. This is because during their growing period, they are not exposed to freezing temperatures.

Sichuan Province is the land of the panda and bamboo. This vegetable grows in the mountains there, and it is huge and different from any grown in the other province. The people there have found a sheath-like membrane as soft as velvet that grows inside the bamboo segments. Each membrane is very light and when dried they are packaged in a bunch of thirty to forty for market. Such a package weighs less than three ounces. This membrane is called *so sun*. It can be found it in larger Chinese groceries in America. Packages of this membrane are quite expensive. They are easy to use and very popular in vegetarian casseroles and soups.

Bamboo Shoots are a species of the genus *Phyllostachys* or Bamboosa. Analysis of many of them show they do not differ much one from another. Their average nutrient analysis shows them to contain ninety-one percent water. They provide only twenty-seven calories per one hundred grams (about three and a half ounces). Like other shoots, those from the bamboo contain vitamin B. They are used in Chinese medicine.

Heart of Palm

Heart of palm is a type of vegetable that is harvested from the inner core of certain species of palm trees. When harvesting heart of palm, the tree is cut down and the bark and fibers are removed, leaving only the heart. Although they're produced in many different areas, most fresh hearts of palm in the United States are actually imported from Costa Rica.

The hearts of palm taste is often compared to artichokes and described as light, mild and crunchy. They look similar to white asparagus and can be baked, blanched, sautéed, marinated or enjoyed straight out of the can.

Heart of palm is incredibly versatile and often used as a meat substitute for those on a vegan or vegetarian diet. It's also revered for its health-promoting properties and is especially rich in fiber, protein, manganese, iron and vitamin C.

Production

Heart of palm is classified in the groups of fresh, prepared and canned processed vegetables. It is usually consumed after being processed and packaged; in other words, it requires a processing stage that adds value. However, heart of palm is also a wonderful fresh vegetable since it maintains its flavor and appearance for two weeks, provided that it is wrapped in plastic and refrigerated immediately after harvesting.

A young stem of heart of palm has three edible parts, of which only one, the heart, accounts for 99% of world consumption. The heart is composed of several spearshaped leaves in development; they are very delicate and are found within the leaves that envelop the stem. The delicate stem, which is 10 to 20 cm long, accounts for only 1% of world consumption, and it is marketed in the form of canned pieces. Even though this stem is difficult to handle during processing because it is not protected by the veins of the leaves, it is a very high-quality product, similar to the heart. Finally, the edible leaves, which are generally discarded as waste in Latin America, are received enthusiastically by chefs in other places such as Hawaii.

Heart of palm usually weighs between 5 and 25 lbs. Its roots are semi-superficial and of a woody consistency, so any damage to the plant tends to be permanent because there is no regeneration. Within non-traditional commodity sales on the international market, heart of palm only has a share of about 2%. However, heart of palm production has increased considerably over the last six years.

Within non-traditional commodity sales on the international market, heart of palm only has a share of about 2%. However, heart of palm production has increased considerably over the last six years.

Areas of Cultivation

The optimal climatological and agroecological characteristics for cultivating heart of palm are as follows:

- Ecological zone: Humid and very humid tropical, according to the Holridge classification.

- Altitude: 0-1000 meters above sea level.

- Luminosity: Requires full exposure to light at least three hours a day in order to get an early start on production.

- Average annual temperature: 24°C to 28°C.

- Precipitation: Cumulative precipitation of 2000 to 4000 mm.

- Relative humidity: 80% or higher.

- Soil: Deep, smooth topography; average texture; sandy, permeable structure to facilitate drainage, since heart of palm is susceptible to excess water. It does not tolerate shallow water levels but is resistant to slightly acidic conditions. If the soil does not have these characteristics, it is necessary to apply fertilizer. The microrhizoids or crimps1 associated with the root system allow the crop to use phosphorus in the acidic soils of the Amazon Region, and it is for that reason that the ground should not be burned before definitive planting.

In Ecuador the heart of palm growers are found in the following zones, which have characteristics suitable for its cultivation:

- Esmeraldas: Quinindé, La Concordia, San Lorenzo, Cayapas.

- Pichincha: Pedro Vicente Maldonado, Santo Domingo, Puerto Quito.

- Manabí: Nueva Delicia.

- Morona Santiago: Yaupi.

- Pastaza: Sarayacu, Teniente Hugo Ortiz.

- Napo: Loreto, Coca, Nueva Rocafuerte.

- Sucumbíos: Nueva Loja, Shushufindi.

These zones have a hot, humid, tropical climate that favors plant growth. Therefore the farms or plantations on which heart of palm are grown often border farms or plantations that produce African palm, bananas, rubber, and macadamia nuts, among other products.

According to climatological studies carried out in the northeastern part of the Amazon Region, the areas that have suitable precipitation, temperature and sunlight conditions for growing heart of palm are the colonization areas of Lago Agrio, Shushufindi and Sacha.

As for topography and soil fertility, Shushufindi and La Joya de los Sachas offer the most suitable conditions for growing heart of palm. They are followed by the areas of Payamino and Coca, which require a larger investment in soil fertility management.

Surface Area and Yield

According to the study on the heart of palm agroindustry in the Ecuadorian economy, over the last six years the cultivated surface of heart of palm has increased at an average annual rate of 90.11%. It is estimated that there are about 4000 hectares of heart of palm planted in Ecuador, of which 1000 were incorporated in 1996. This addition called for an investment of approximately 1.5 million USD. Of these 4000 hectares, 3200 are on the coast, and the other 800 are in the Amazon Region, mainly in Napo and Sucumbíos.

The improvement in the price of each stem of heart of palm led to a spike in planting. This price improvement was due to better marketing and a contraction in the supply of Costa Rican heart of palm.

Varieties

It is important to point out that heart of palm can be found in the wild, where most of the palms have only one stalk or stem and therefore die if they are cut. This is the case of the Euterpes heart of palm, a very widespread wild species in Brazil. However, this does not occur with cultivated species such as the Bactris gasipaes.

The study "Introduction and Evaluation of Pejibaye for Heart of Palm Production in Hawaii" done by Purdue University cites the three species that are listed below as the most commonly cultivated:

- E edulis: Should be planted and remain in slight to moderate shade for three to five years. The plants of this species have a high degree of mortality during the crop formation stage. In high-density plantations, production is excellent, but the hearts of palm are small. Inversely, in less dense plantations production is smaller but the hearts of palm are larger. Harvesting begins six to eight years after planting and, due to the fact that these palm trees have only one stem, the plantation must be fully replanted. The productivity of crops that combine plants of different ages is, in the best of cases, barely half that of a low-density crop.

- E oleracea: Grows faster than E edulis, since it reaches harvesting size between four and six years after planting, both in its natural ecosystem and under artificial conditions. This species should be planted under slight shade, which can be eliminated after one year. In comparison to E edulis, plant mortality during crop formation is low in humid areas and high in areas of the Amazon plains. Productivity is similar to that of the E edulis and shows similar trends with respect to density. Due to the fact that it is a palm tree of the "caespitose" type, each group of plants produces another stem of heart of palm after 18 to 24 months after planting, thus enabling production to be continuous.

  It is estimated that after the first harvest the productivity of one hectare of E oleracea is 1.4 times greater.

- Pejibaye or Chontaduro: This variety can be planted in full sunlight and needs slight shade during the period of seedling development. If managed well, it does not have from a high degree of mortality during the first years of life. Under agricultural conditions similar to those of its natural ecosystem, the palm tree grows fast and responds positively to the application of fertilizers and other inputs, to reach its ideal harvesting size 18 to 30 months after planting. Just as the E oleracea, the pejibaye or chontaduro is a caespitose, which lends itself to continuous production. Each group of plants produces a new heart every 9 to 15 months. In Costa Rica, the production of this variety doubles after the first harvest.

Cultivation Season

The heart of palm is cultivated throughout the year in a scaled manner, obtaining up to two harvests per plant per year.

Value Chain

Cultivation

About 5000 seedlings are planted on one hectare; however, when these are initially kept in a nursery, it is necessary to count on an additional 10% in order to replace those that die in the nursery itself or when transplanted to firm soil.

There are basically two planting methods:

1. In beds, where the seeds are planted beside each other in rows, with bare roots. This method is not suitable because it leads to a greater mortality index during transplantation to firm soil. The pioneers of heart of palm cultivation used this method, and the plant mortality index was as high as 40%.

2. In plastic bags, where each seed is planted individually within a plastic bag that protects its roots. The growers that have learned from the mistakes of the early palm growers have managed to reduce plant mortality to about 10%.

In order to maintain good plantation yields, it is necessary to have an irrigation system in order to have uniform moisture distribution because there are significant dry periods that reduce crop productivity. Furthermore, if there is excessive moisture, the crop becomes highly productive, but its quality is diminished. In order to obtain a high-quality product, it is important to have a suitable infrastructure. The figures considered under the item of civil works for a heart of palm plantation include the following: ranch house and offices, guard houses, fences, roadways and drainage, irrigation pipes, greenhouses, and sheds or storehouses.

Labor for heart of palm crops is important; its intervention in both production and processing is decisive. For this reason, constant advising from an expert on heart of palm cultivation is important, in addition to training the workers directly involved in the agroindustrial process.

Most of the heart of palm growing farms in Ecuador have signed an agreement with the heart of palm processing plants, whereby they commit to selling their product to a specific processor in exchange for certified seed and technical assistance. During the crop development stage, the technical assistant works 16 hours per month on the farm at a price of two dollars per month per hectare. The technical assistance consists of training the farm workers in sound crop management and ensuring a high-quality product in accordance with the requirements of the processors, whose guidelines are based on market demands.

The different agricultural stages of heart of palm before the industrial processing stage are as follows:

a. *Seed selection*

The best lines of stools are selected from a germ plasm bank, especially those favored by genetic variability, which increases resistance to pests and diseases. The best plants are the ones that are free of pests and diseases, with a high and stable production of sprouts and with long joints. The selection process is important both for new planting and for replanting.

In the production period, clusters with healthy fruit that has reached complete physiological maturity must be collected. In other words, the fruit should be red, orange or yellow, depending on the variety. Once the clusters have been harvested, the fruit must be classified according to their stage of maturity, separating them into ripe, colored and green. The green and overripe fruit is not used.

The seed from ripe fruit germinates first, whereas the seed from the colored fruit takes two months. Once the seed has been obtained, the pulp is removed from it, and it is soaked, washed, air-dried and disinfected. The clusters should not be exposed to the sun because the increase in temperature can lead to a fermentation process and the death of the embryo.

b. *Pulp removal*

The pulp should be removed on the day of harvesting or immediately afterwards, preferably manually, using a knife to peel off the mesocarpium. A manual pulp remover can also be used, taking care not to harm the seed. The fruit should not be stepped on, nor hit with sticks or stones because that causes lesions in the seed cover, which permits the entry of fungus and possible decomposition.

c. *Soaking*

The depulped seed should be soaked in suitable recipients for 48 to 72 hours so that it can reach the point of maximum absorption. The floating seeds should be discarded because they are empty.

d. *Washing and air-drying*

In order to loosen any remaining pulp, the seed should be washed immediately after the soaking process, until it is totally clean, in order to avoid the proliferation of fungus. It should then immediately be treated with 1% sodium hypochlorite or disinfected with some type of fungicide. It should then be placed in a well ventilated area, on mesh, newspaper, or a clean wooden or cement floor so that the seed can eliminate the free water on its surface. It is important not to expose the seed to the sun nor to excess air, in order to avoid its desiccation. It should not be stored for very long either, and it is preferable for it to be planted immediately after its disinfection in the send seed beds, in rows 5 cm apart and 3 to 5 cm deep.

Planting is done in seed beds on easily accessible, flat terrains, near a water source and the final planting areas and free of rocks, clumps, sticks, weeds, pests and pathogens. The seed beds are usually protected by a roof or shelter, whose dimensions will depend on the number of seeds to be planted. A shelter 1.5 m wide and 30 m long can hold up to 150,000 seedlings. It is important for the cover to keep 60% to 70% of the volume in the shade; therefore, it should be approximately 2.5 m high. In order to avoid the entry of animals, the shelter can be protected on the sides using wire mesh, plastic or sheets of zinc.

The seed beds are temporary, and the time period in which they are built depends on each region. In the Amazon Region, for example, the seed is produced in the months of February to May. The seed begins to germinate 40 days after planting, and it takes 60 to 120 days to complete the process.

e. Transplanting *to nurseries or seedling sheets*

The germinated seeds that present plumes are ready to be transplanted into the nursery bags, which can hold between 1 and 2 kg of soil. Another option is to transplant the germinated seeds directly onto sheets measuring 1 x 30 m and prepared previously with loose soil or sand. In order to avoid the effects of possible flooding, the sheets should be built on the surface of the soil and be protected by a drainage system. The seedlings remain on the sheets for four to six months, planted 10 to 20 cm apart.

f. *Definitive planting*

When the seedlings have at least six leaves and are about 25 cm high, they are definitively planted in holes that are 20 cm wide x 20 cm long x 20 cm deep, at a distance of 2 x 1 m between aisles and plants, respectively. This yields a density of 5000 plants/ha, which increases to 10,000 plants/ha after two years since each plant produces sprouts.

g. *Fertilization*

The following chart indicates the amounts of nutrients that this crop requires for optimal development and production. If the soils do not have suitable conditions, a fertilization program can be designed on the basis of the following requirements:

Nutrients	Nutrient removal by area of cultivation Kg/ha	Nutrient removal by gross cultivation Kg/ha
Nitrogen	531.00	28.00
Phosphorus	37.90	4.80
Potassium	248.30	31.00
Calcium	64.80	4.7
Magnesium	43.00	3.90
Iron	1.83	0.03
Copper	0.18	0.02
Zinc	0.25	0.05
Manganese	2.27	0.08
Sulfur	47.23	3.36
Boron	0.56	0.03

h. *Weed control*

Weed control prevents competition for nutrients and facilitates fertilizer application. It can be done through manual clearing, provided that damage to surface roots is avoided. The number of weedings per year depends on soil conditions. It is estimated that four or five weedings are needed in the first year, and three from the second year on. In sectors in which labor is scarce or expensive, herbicides can be used to remove weeds. One way to prevent weed growth is to spread plant matter between the rows of heart of palm seedlings.

i. *Sprout control*

It is necessary to eliminate the sprouts that grow out of the aerial part of the stem due to the fact that they are not attached to the rhizome and do not contribute roots. It is also possible to remove the ones that emerge towards the adjacent rows, in order to direct the sprouts in the direction of the row in which they are found. An average of four sprouts larger than 30 cm may be maintained for each plant.

j. *Pest and disease control*

In nurseries and plantations, the most common diseases are Pestalotia, Cescóspora, Mycosphaerella spp, Black Spot, and Fusariun moniliforme, which causes the disease known as "flecha." All of these affect the foliage, causing spots and sometimes death. In most cases, the presence of these diseases is due to inadequate nutrition or drainage. It is possible to control them in nurseries using fungicides.

k. Cutting

Within a year to a year and a half after definitive planting, heart of palm cutting begins. The stems to be cut should have a base diameter of between 12 and 15 cm. The cutting is done with a machete, 40 cm above the ground. In the case of ecotypes that have protuberances on the stem, the cut is made below the protuberance.

Leaves are taken off the harvested stems, leaving only the internal layers that envelop the heart of palm, in order to protect it against the entry of microorganisms and avoid breakage. The end parts that will not be transported to the factory are also removed. The stems, approximately 70 cm long, should remain in the shade and should be transported to the processing plant the same day that they are cut.

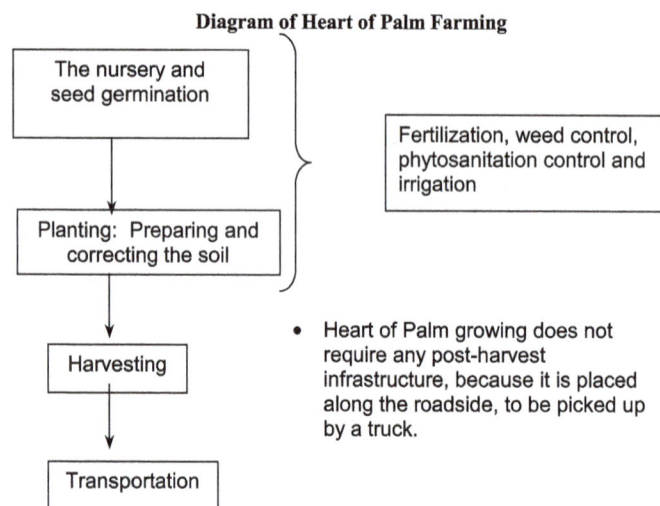

Diagram of Heart of Palm Farming

```
┌──────────────────────┐
│ The nursery and      │
│ seed germination     │          ┌───────────────────────────┐
└──────────┬───────────┘          │ Fertilization, weed control,│
           │                      │ phytosanitation control and │
           ▼                      │ irrigation                  │
┌──────────────────────┐         └───────────────────────────┘
│ Planting:  Preparing and │
│ correcting the soil      │
└──────────┬───────────┘
           │                    • Heart of Palm growing does not
           ▼                      require any post-harvest
┌──────────────────────┐          infrastructure, because it is placed
│ Harvesting           │          along the roadside, to be picked up
└──────────┬───────────┘          by a truck.
           │
           ▼
┌──────────────────────┐
│ Transportation       │
└──────────────────────┘
```

Processing

Series production is the most suitable for processing heart of palm, since it is homogeneous mass production. In addition to simple technology, heart of palm processing requires labor trained in the processes of selecting, cutting, peeling and packaging the product.

The different stages of processing heart of palm prior to its points of sale on the market are described below.

a. Reception and peeling of raw materials

The raw material arrives at the heart of palm processors in unrefrigerated trucks. Then the stems are counted and inspected for freshness, using their color and texture as a reference. Then the selected raw material is stored, without refrigeration, for its later placement in the hoppers that feed the peeling line.

The peeling line consists of a three-level band. The raw material is placed on the intermediate level for manipulation by the operators that peel the stems and separate the outside part from the inside part. They place the waste on the lowest band and the heart on the highest.

The waste accumulates at the end of the band and is removed for its later delivery to companies that use the waste material to produce earthworm humus. Meanwhile, the heart of palm is placed in tubs containing 70 to 80 liters of water, in order to avoid product oxidation or mistreatment, and so that the bark residues still adhering to the heart will loosen.

The heart of palm is then distributed in tubs containing 7 to 8 liters of water. This is reused approximately six times before being deposited in a sump that retains the thick solids. Every time the water is changed, the tubs are washed with a ½ hose with an output of approximately 5 to 6 liters per minute. Washing the tubs takes one to two minutes, and about four tubs are used each hour. The areas devoted to receiving and peeling the heart of palm are cleaned without water.

b. Cutting and sorting

After the peeling operation, the heart of palm enters two lines for sorting and cutting. These lines have two wash tubs that make it possible to eliminate microorganisms by applying a 35 m³ per day flow. The hearts of palm are cut on tables that facilitate fragmentation in pieces 10 cm long.

As the pieces of heart of palm submerge in the tub, the first selection is made, using hardness as the criterion. Later, in the wash tub a second selection is made using the same criterion. The soft heart of palm goes directly to a canning stage and the hard pieces are subject to a precooking operation. The washing operation is done at intervals of 30 minutes, which makes it possible to evacuate the used water in the drainage line and substitute fresh washwater.

c. *Packaging*

Once selected and cut, the heart of palm is canned manually. If hard pieces are found, they are removed and sent to the scalding process. The filled cans are placed on a table with raised edges, which allows water from the runoff from the previous stage to accumulate.

d. *Weighing*

Each can is weighed manually, using two scales with a precision of ± 5 g. The weight is adjusted using cooked or raw pieces of heart of palm.

e. *Addition of filler liquid*

The cans are placed on a conveyor belt and passed through the apparatus that dispenses filler liquid, which contains citric acid, salt and sugar. The filler liquid remains at a temperature of 85°C. Any excess liquid goes into a tub from which it is reintroduced into the circulation system for canning.

f. *Removal of excess liquid*

The cans go to a transport tunnel where the excess filler liquid spills onto the floor.

g. *Sealing*

In this stage the cans containing the heart of palm are sealed mechanically. A single operator performs this activity. The cans are placed in metal baskets and then transported to the pasteurization area. During sealing, no water is added to the product, and no water is used in the process Small amounts of liquid do leak out of the cans, but they are estimated as no more than 10 l/day.

h. *Pasteurization (heat treatment)*

Heating: During this stage the metal basket containing the sealed cans is introduced for 22 minutes into an airtight autoclave or pressure pot that can reach a temperature of 104 °C (steam-heated). Once the inside temperature of the autoclave drops to 90 °C, it is opened and the basket containing the cans is removed. There are four autoclaves on the production line, each one capable of holding 120 one-kg cans. During the heating process the autoclave emits steam directly into the workspace. The discharge volumes are not known.

Cooling: This process consists of introducing the metal basket into a cold-water bath, in order to produce a thermal shock and thus guarantee the sterilization of the final product. The cooling takes place in a 4.4m tank. In order to guarantee the cooling temperature, the system is continuously supplied with fresh water. Flow rates vary, and they are controlled by an operator. Approximately 10 of water are used daily in the cooling process. No kind of contamination is apparent.

Flow Diagram of the Industrial Production of Canned Heart of Palm

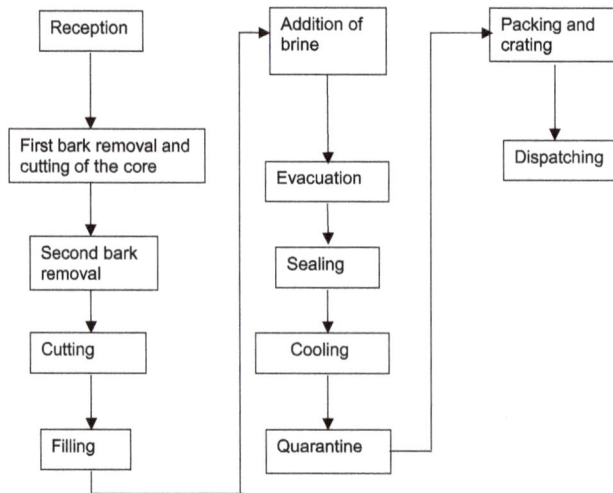

Packing and Storage

Heart of palm is packed in cans or glass jars. The cans are made of thin steel covered with layers of sanitary enamel, and the glass jars are usually round and are covered with an enameled lid. Vacuum seals are a third packaging alternative. If the heart of palm is to be consumed as a fresh vegetable, it is sufficient to cover it with a thin layer of plastic wrap immediately after harvesting.

It should be mentioned that marketing of peeled heart of palm, packaged in plastic bags that are kept refrigerated, has been very successful. The most common presentations are 220, 235, 250 y 500 net g of drained heart of palm. The boxes of 12 gross kg contain approximately 10.2 net kg of product, including water, which is equivalent to 6 net kg of drained heart of palm. Ecuadorian heart of palm is marketed in a brine solution, with no chemicals or artificial preservatives.

Classification System

Container	Thickness	No. of hearts per container
Large can	Thin	13 to 19
	Medium	8 to 12
	Thick	4 to 7
Small can	Thin	6 to 12
Jar	Medium	7 to 9
	Thick	4 to 6

Container	Net Weight	Drained Wt.	Containers per box	Boxes per container
Large can	800 g.	500 g.	12	1800
Small can	400 g.	235 g.	24	1500
Jar	420 g.	250 g.	12	2250

Benefits of Hearts of Palm

Promote Digestive Health

Fiber is important to many components of health, particularly when it comes to digestion. It moves slowly through the body undigested, adding bulk to stool and preventing issues like constipation. Fiber also acts as a prebiotic to promote the growth of beneficial bacteria in the gut. Your gut microbiome plays a central role in health and disease and has even been linked to obesity, immunity and mental health.

Hearts of palm are an excellent source of fiber, packing 3.5 grams of fiber into each cup. That means that adding just a single cup of hearts of palm into your diet can knock out up to 14 percent of some people's fiber needs for the whole day.

Aid in Weight Loss

High in both protein and fiber yet low in calories, hearts of palm make a great addition to the diet if you're looking to lose weigh fast. Some studies show that protein helps reduce levels of ghrelin, the hunger hormone, to ward off cravings and decrease appetite. Meanwhile, fiber keeps you feeling full to promote satiety and reduce intake.

Because of their unique taste and texture, hearts of palm are often used as a vegan meat alternative in many recipes. Try subbing them into your next salad or sandwich in place of meat to cut down on calories and fat and help keep your weight in check.

Support Bone Health

Osteoporosis is a common concern as you start to get older and begin to lose bone mass. In fact, approximately 1.5 million Americans experience fractures due to bone disease each year, and by 2020, it's projected that one in two adults over 50 will have or be at risk for developing osteoporosis of the hip.

Hearts of palm are loaded with manganese, a mineral that's key to bone health. A deficiency in this crucial nutrient can lead to alterations in bone metabolism and a decrease in the synthesis of bone tissue. According to an animal study out of Sookmyung Women's University's Department of Food and Nutrition in Seoul, South Korea, supplementation with manganese for 12 weeks was even able to increase bone formation and bone mineral density in rats.

Stabilize Blood Sugar

Maintaining high blood sugar for long periods of time can come with some serious side effects, including nerve damage, an increased risk of infections and even kidney damage.

Thanks to its content of both fiber and manganese, heart of palm can help you maintain normal blood sugar to sidestep negative symptoms. Fiber slows the absorption of sugar in the bloodstream to prevent spikes and crashes in blood sugar levels. Manganese may also play a role in blood sugar control, with some animal studies suggesting that a deficiency in manganese could impair insulin secretion and carbohydrate metabolism.

Help Prevent Anemia

Anemia is a condition characterized by a lack of healthy red blood cells in the body, causing a long list of possible anemia symptoms like fatigue, light-headedness and brain fog. Although there are a number of factors that can contribute to anemia, one of the most common causes is a deficiency in certain nutrients like iron.

One cup of hearts of palm contains 25 percent of the iron you need in a day, which can help you easily meet your needs to prevent conditions like iron-deficiency anemia. Not only that, but it also contains a good chunk of vitamin C, helping enhance iron absorption even more.

Boost Immunity

Hearts of palm are rich in many important vitamins and minerals that are essential for keeping your immune system running smoothly. Vitamin C, zinc and manganese, in particular, are all vital for warding off infections and disease to promote better health.

For example, one review published in the Annals of Nutrition concluded that getting enough vitamin C and zinc can reduce symptoms and shorten the duration of respiratory tract infections, plus improve the outcomes for conditions like pneumonia, malaria and diarrhea. Manganese, on the other hand, protects against oxidative stress and fights off free radicals that can contribute to chronic disease

Inflorescence Vegetables

Inflorescent Vegetables are typically vegetables with an edible flower. They are comprised of clusters of flowers arranged on a stem containing a main branch or complicated arrangement of branches including flowers, flower buds, and their associated stems and leaves.

Generally all parts of the vegetable are edible. Inflorescent vegetables are packed with nutritious value and often the choice of vegetables for dieters as they have little caloric value but are filling and hearty.

Use in casseroles, steamed as a side dish, raw in salads, pureed and made into soups and sauces.

Artichoke

The artichoke (Cynara scolymus), is as old as human civilization itself; artichokes were cultivated by the ancient Greeks, Romans and Egyptians. It is native to the Mediterranean region, an area of the world with one of the lowest rates of chronic disease and one of the highest life expectancies. It is a perennial in the thistle group of the sunflower (Compositae) family. The "vegetable" that we eat is actually the plant's flower bud. If allowed to flower, the plant will produce a beautiful blossom of violet-blue color.

Here in the United States, chances are the artichoke you bring home from the market was grown in California. Nearly 100% of all artichokes grown commercially in the United States are grown in California and 80% of those in and around the town of Castroville, in Monterey County, making it the artichoke.

"heart" of the world

Despite it's long history, if you consider the lowly artichoke nestled in its market display and shielded by it's thorny exterior, it's ultimate value may not be immediately obvious. In this era of fast food, the artichoke remains defiant—it offers no concessions to those who want a quick meal. It requires time to prepare and time to eat, petal by petal, until at last you reach the delectable heart inside.

But even though it looks more like a hand grenade than a vegetable, it would be a mistake to ignore the many health benefits that artichoke has to offer.

In a 2004 study conducted by the U.S. Dept. of Agriculture of 100 common foods found in U.S. markets, it was determined that cooked artichoke had an extremely high oxygen radical absorbance capacity, compared to other vegetables; it's total antioxidant capacity was surpassed only by dried pinto and kidney beans. This is important because artichoke can help to eliminate free radicals from the body, thereby reducing the oxidative stress that has been associated with the development of many chronic and degenerative diseases, including cancer, heart disease and neural degeneration.

In addition to being a powerful antioxidant, the artichoke is also a choleretic, i.e., it helps to stimulate the production of bile by the liver. Artichokes also act as hepato-protectors; protecting liver cells from damage that can occur as they process toxins in the body. A 2003 study tested four commercially prepared whole artichoke extracts and confirmed that, when properly prepared and constituted and with the correct dosage, artichoke extract clearly demonstrated these beneficial effects on the liver.

Contraindications for the use of artichoke would include any form of bile duct blockage or obstruction and those with gallstones should seek the advice of their physician. People with an allergy to plants of the Asteraceae family should avoidartichoke; contact with the leaves or the plant itself may cause dermatitis in sensitive individuals.

Fortunately, those who are able to consume artichokes can enjoy them in many ways, not only petal by petal. The benefits of artichoke may also be enjoyed in the form of a liquid extract, enabling you to drink your "artichoke a day".

Agricultural Output

Artichoke head with flower in bloom

Artichokes for sale

Today, cultivation of the globe artichoke is concentrated in the countries bordering the Mediterranean basin. The main European producers are Italy, Spain, and France and the main American producers are Argentina, Peru and the United States. In the United States, California provides nearly 100% of the U.S. crop, and about 80% of that is grown in Monterey County; there, Castroville proclaims itself to be "The Artichoke Center of the World", and holds the annual Castroville Artichoke Festival. Most recently, artichokes have been grown in South Africa in a small town called Parys located along the Vaal River.

Top 12 globe artichoke producers in 2014		
Country	Production (tonnes)	Footnote
Italy	451,461	
Egypt	266,196	
Spain	234,091	
Argentina	105,236	Im
Peru	103,348	
Algeria	81,106	
China	78,055	Im
Morocco	55,234	
United States	43,050	
France	38,354	
Turkey	34,576	
Tunisia	19,000	
World	**1,573,363**	**A**

* = Unofficial figure | [ ] = Official data | A = May include official, semi-official or estimated data
F = FAO estimate | Im = FAO data based on imputation methodology | M = Data not available

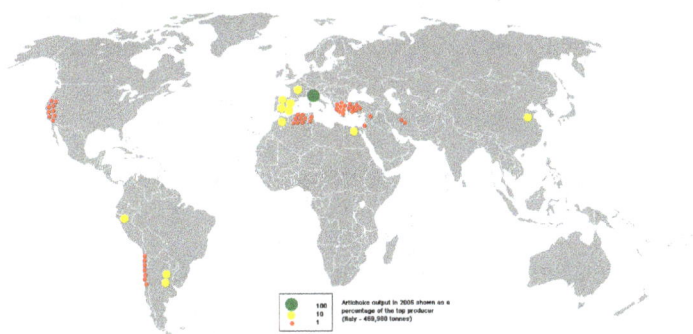

Artichoke output

Artichokes can be produced from seeds or from vegetative means such as division, root cuttings, or micropropagation. Although technically perennials that normally produce the edible flower during only the second and subsequent years, certain varieties of artichokes can be grown from seed as annuals, producing a limited harvest at the end of the first growing season, even in regions where the plants are not normally winter-hardy. This means home gardeners in northern regions can attempt to produce a crop without the need to overwinter plants with special treatment or protection.

The seed cultivar 'Imperial Star' has been bred to produce in the first year without such measures. An even newer cultivar, 'Northern Star', is said to be able to overwinter in more northerly climates, and readily survives subzero temperatures.

Commercial culture is limited to warm areas in USDA hardiness zone 7 and above. It requires good soil, regular watering and feeding, and frost protection in winter. Rooted suckers can be planted each year, so mature specimens can be disposed of after a few years, as each individual plant lives only a few years. The peak season for artichoke harvesting is the spring, but they can continue to be harvested throughout the summer, with another peak period in midautumn.

When harvested, they are cut from the plant so as to leave an inch or two of stem. Artichokes possess good keeping qualities, frequently remaining quite fresh for two weeks or longer under average retail conditions.

Apart from food use, the globe artichoke is also an attractive plant for its bright floral display, sometimes grown in herbaceous borders for its bold foliage and large, purple flower heads.

Varieties

Some varieties of artichoke display purple coloration

Traditional Cultivars (Vegetative Propagation)

- Green, big: 'Vert de Laon' (France), 'Camus de Bretagne', 'Castel' (France), 'Green Globe' (USA, South Africa).

- Green, medium-size: 'Verde Palermo' (Sicily, Italy), 'Blanca de Tudela' (Spain), 'Argentina', 'Española' (Chile), 'Blanc d'Oran' (Algeria), 'Sakiz', 'Bayrampasha' (Turkey).

- Purple, big: 'Romanesco', 'C3' (Italy).

- Purple, medium-size: 'Violet de Provence' (France), 'Brindisino', 'Catanese', 'Niscemese' (Sicily), 'Violet d'Algerie' (Algeria), 'Baladi' (Egypt), 'Ñato' (Argentina), 'Violetta di Chioggia' (Italy).

- Spined: 'Spinoso Sardo e Ingauno' (Sardinia, Italy), 'Criolla' (Peru).

- White, in some places of the world.

Cultivars Propagated by Seeds

- For industry: 'Madrigal', 'Lorca', 'A-106', 'Imperial Star'.

- Green: 'Symphony' 'Harmony'.

- Purple: 'Concerto', 'Opal', 'Tempo'.

Broccolini

Broccolini is a member of the Brassica family, alongside broccoli, cauliflower, and cabbage. It was invented many years ago in Japan where, using plant-breeding techniques, broccoli and Chinese kale were combined to create a more flavorsome Brassica. Looking similar to broccoli, this vegetable goes by many names, including tenderstems, sweet baby broccoli, asparation, bimi, broccoletti, and Italian sprouting broccoli. It is also sometimes referred to as baby broccoli.

Broccolini Appearance and Taste

Broccolini is similar in appearance to broccoli in that it is made up of a green stem topped with florets. But whereas the broccoli stem can be very thick and tough, the stem of broccolini is thin and tender. And instead of densely packed florets, broccolini has looser crowns that seem more leaf-like.

Broccolini has a mild, somewhat sweet, distinctive flavor and texture more like asparagus than traditional broccoli. It is tender from floret to stem so you can eat the whole vegetable. This is unlike ordinary broccoli which tends to have a sometimes woody stem.

Broccolini Nutritional Value

Broccolini is considered a superfood since it is rich in vitamin C, providing 100 percent of the daily requirement. It also contains calcium, vitamins A and E, potassium, folate, and iron. Pair this with only 35 calories per serving and you've got one healthy vegetable.

Production

Climate

Broccolini grows in cool climates and is intolerant to extreme climates. It is more sensitive to cold temperatures than broccoli but less sensitive to hot temperatures.

Growth and Distribution

Broccolini takes 50–60 days to grow after being transplanted. It is harvested when the heads are fully developed but are not flowering. By cutting off the head, the harvest time is extended by four weeks as new shoots of smaller heads now grow. After being harvested, the plant is cooled to 0°C, preventing the flower heads developing.

Broccolini is grown near the central California coast during the spring, summer, and fall seasons and Yuma, Arizona throughout the winter. It is sold throughout the United States, Canada, and the United Kingdom, as well as 5 states in Australia.

Broccoflower

The term broccoflower is actually used to describe two different, though similar, vegetables. The name comes from a cross between broccoli and cauliflower, two vegetables that are actually so closely related that they can be easily cross-pollinated.

One type of broccoflower looks just like white cauliflower, but is lime-green in color. There are many different kinds of green cauliflower, and several are the result of the cross-pollination of broccoli and cauliflower.

Romanesco Broccoli

Another vegetable that often goes by this name is Romanesco broccoli, an unusual green vegetable that has a unique fractal pattern, resembling tiny pine trees, on its head.

Both types of broccoflower are generally milder, more tender, and slightly sweeter than either broccoli or caulifower. For those reasons, they are delicious raw, and make a great, conversation-starting addition to crudité platters.

Broccoflower can be substituted for cauliflower or broccoli in any recipe that calls for them. Beware of overcooking. Just like broccoli, it can become stringy and unpleasant when overcooked.

Health Benefits of Broccoflower

Because broccoflower is a combo of two super healthy veggies, it's high in health benefits, too. In fact, it's full of all the nutrients and antioxidants that you expect from broccoli to boost wellness.

References

- Lim, T. K. (2014). Edible Medicinal And Non-Medicinal Plants: Volume 7, Flowers. Springer. p. 625. ISBN 978-94-007-7394-3

- Stems-foods-that-are-extremely-healthy-to-eat-083795: boldsky.com, Retrieved 06 July 2018

- Piccone, Marie. "Understanding the retail performance of broccolini using a tool for determining in store performance and customer demand" (PDF). AusVeg. Horticulture Australia Ltd. Retrieved 17 September 2016

- Asparagus-production: extension.psu.edu, Retrieved 10 May 2018

- Ceccarelli N., Curadi M., Picciarelli P., Martelloni L., Sbrana C., Giovannetti M. "Globe artichoke as a functional food" Mediterranean Journal of Nutrition and Metabolism 2010 3:3 (197–201)

- How-to-grow-celtuce-or-stem-lettuce: hobbyfarms.com, Retrieved 26 April 2018

- "Sakata Home Grown Presents: Broccolini® Ideas». Sakata Vegetables. 1 January 2013. Retrieved 22 September 2016.

- Artichoke-a-liver-healthy-vegetable: hbmag.com, Retrieved 15 April 2018

- What-is-tenderstem-broccoli-435435: thespruceeats.com, Retrieved 19 June 2018

Common Diseases in Vegetables

The scientific study of diseases in plants that is caused by pathogens or physiological factors is under the science of plant pathology. Bacteria, viruses, fungi, protozoa, etc. can cause diseases in plants. Black dot, beet vascular necrosis, Tobamovirus, Alternaria solani, etc. are some of the common diseases that affect vegetable production. These diseases along with others have been extensively covered in this chapter.

Black Dot

Black dot is a common disease of potato. It is most often observed on tubers but it can affect all parts of the plant. The disease has probably been underestimated in the recent past as the symptoms are similar to more common potato diseases. On potato foliage symptoms are nearly indistinguishable from early blight and on tubers it produces blemishes that are easily mistaken for silver scurf. Although not as serious as other more common potato diseases such as black scurf, silver scurf, or common scab, it can be more devastating as it affects all parts of the plant. Above ground it can infect the vascular system causing wilt, and below ground it can cause severe rotting of roots, shoots and stolons, leading to early plant decline, discolored tubers and reduced yields.

Symptoms

Figure: Black spots on the surface of a black dot infected tuber. These are microsclerotia of the black dot pathogen *Colletotrichum coccodes*

Figure: Black dot mircosclerotia on the surface of a potato root, and root hair

Black dot is named after the abundant black dots that form on tubers. The black dots are microsclerotia that are often just visible to the naked eye. They are not restricted to tubers and can also be found on stolons, roots and stems both above and below ground. Foliar symptoms are not often observed in Michigan, but this may be due to the fact that they bear a resemblance to the small brown to black flecks characteristic of early blight. Stem lesions are more common in Michigan than foliar symptoms, especially towards the end of the growing season. Stem lesions tend to form around the base of leaf petioles. As with leaf lesions they initially start out as small brown flecks. These gradually coalesce forming lesions which may girdle the stem. As lesions mature they develop circular to irregularly shaped, white to straw-colored centers with wide margins that vary in color from brown to black. Microsclerotia form in the center of the lesions and are often clearly visible against the pale background. As infected tissue senesces, microsclerotia may become abundant covering the entire surface of the stem. Microsclerotia often appear at the base of the plant up to several inches above soil level late in the season and after vine kill.

Lesions on below ground stems and stolons may resemble Rhizoctonia lesions but are darker in appearance. Infection of root cortical tissue causes sloughing of the periderm and may result in severe rotting and early plant death. Microsclerotia and mycelia may also be abundant both internally and externally on roots and stolons.

Tuber symptoms appear as a brownish to gray discoloration over a large portion of the tuber or as circular to irregularly shaped areas. Black dot may develop a silvery sheen during storage, which can be confused with silver scurf. However, black dot tends to show much more irregularly shaped patches with less well-defined margins than silver scurf. Inspection with a hand lens (10x) will quickly differentiate the regularly spaced black dots from the bunched threads of silver scurf.

Disease Cycle

Black dot is caused by the fungus Colletotrichum coccodes. The pathogen has a wide host range, occurring on other plants in the Solanaceae family including eggplant, pepper, tomato and weeds such as hairy nightshade (Solanum sarrachoides). Colletotrichum coccodes readily produces microsclerotia on senescing plant tissue and the surface of tubers. These structures allow the pathogen to survive for long periods of time in the soil. In Michigan, the pathogen survives between growing seasons as sclerotia in infested plant debris and soil, on infected potato tubers and in overwintering debris of susceptible solanaceous crops and weeds. In the spring, sclerotia develop into acervuli. These are fruiting bodies which produce masses of spores. Spores serve as the primary inocula to initiate disease. Initial inoculum is readily moved within and between fields, as the spores are easily carried by air currents, windblown soil particles, splashing rain and irrigation water. Poor soil drainage and low plant fertility are thought to increase disease incidence and severity.

Figure: Tuber symptoms appear as a brownish gray discoloration over a large portion of the tuber

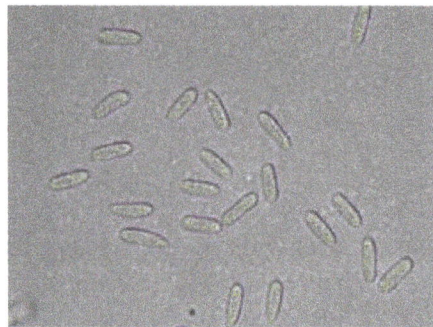

Figure: Spores of the black dot pathogen *Colletotrichum coccodes*

Spores of C. coccodes are produced on potato plants and plant debris between 45° and 95° F. However, in greenhouse studies limited infection occurred below 59° F. As with most Colletotrichum species, C. coccodes favors temperatures above 68° F, and free moisture from rain, irrigation, fog or dew is required for spore germination and infection of plant tissues. Spores landing on susceptible plant tissue germinate and may penetrate tissues directly through the epidermis and or through wounds such as those caused by mechanical injury or insect feeding. On tomato leaves, C. coccodes has been reported to colonize lesions caused by Alternaria solani (early blight) and flea beetles.

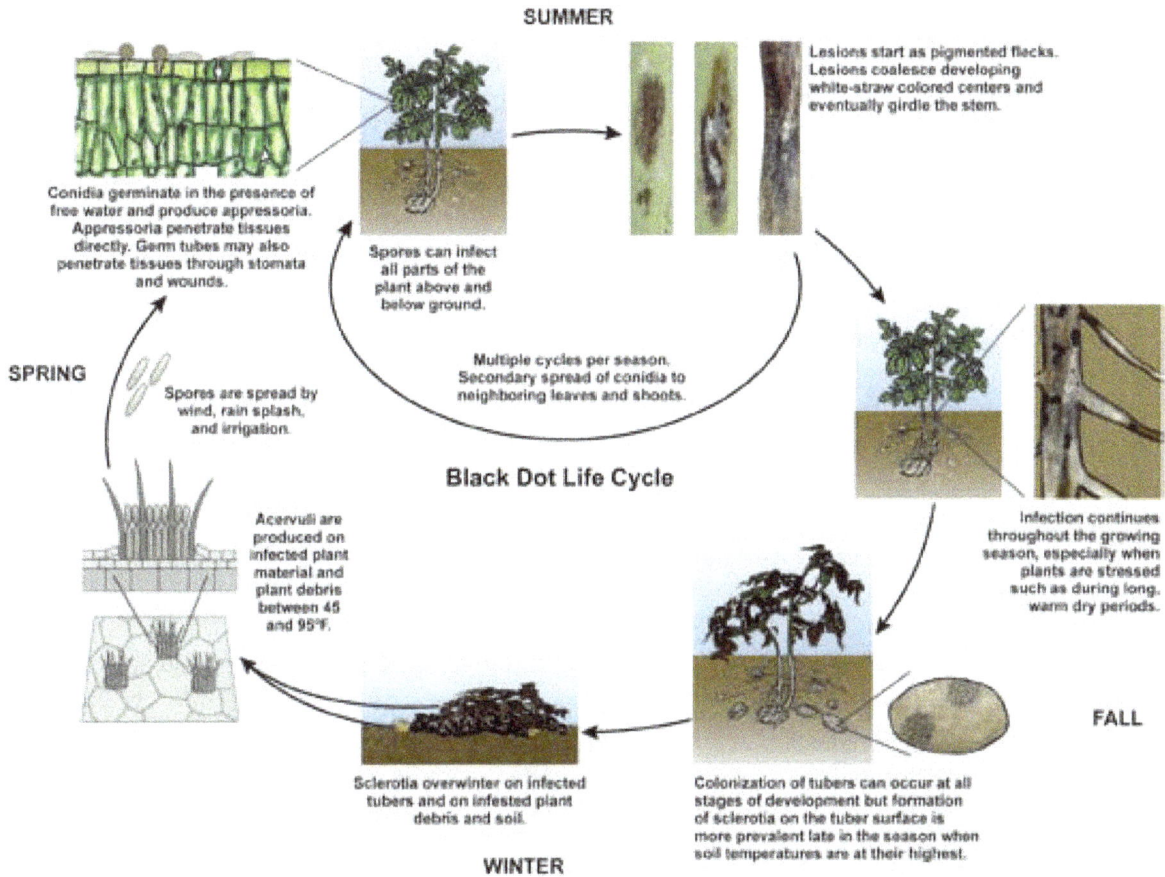

SUMMER

Conidia germinate in the presence of free water and produce appressoria. Appressoria penetrate tissues directly. Germ tubes may also penetrate tissues through stomata and wounds.

Lesions start as pigmented flecks. Lesions coalesce developing white-straw colored centers and eventually girdle the stem.

Spores can infect all parts of the plant above and below ground.

Multiple cycles per season. Secondary spread of conidia to neighboring leaves and shoots.

SPRING

Spores are spread by wind, rain splash, and irrigation.

Black Dot Life Cycle

Acervuli are produced on infected plant material and plant debris between 45 and 95°F.

Infection continues throughout the growing season, especially when plants are stressed such as during long, warm dry periods.

FALL

Sclerotia overwinter on infected tubers and on infested plant debris and soil.

Colonization of tubers can occur at all stages of development but formation of sclerotia on the tuber surface is more prevalent late in the season when soil temperatures are at their highest.

WINTER

Figure: The disease cycle of the black dot pathogen, Colletotrichum coccodes

Infection of below-ground plant parts continues throughout the growing season, especially when plants are under stress, such as during long, warm dry periods. Production of microsclerotia is greatest at high temperatures and increasing disease incidence has been associated with increasing soil temperatures. Colonization of tubers can occur at all stages of development, but formation of sclerotia on the tuber surface is more prevalent late in the season when soil temperatures are declining from the peak in high summer.

Monitoring and Control

Effective management of black dot requires implementation of an integrated disease management approach. The disease is controlled primarily through the use of cultural practices and fungicides.

Cultural Control

One of the most important approaches to black dot control is to reduce the amount of inoculum in soil through the use of cultural practices such as crop rotation, removal of crop debris, volunteer and cull potatoes from the field, and eradication of weed hosts. Since black dot microsclerotia can persist in the field for up to 2 years, a 3 to 4 year rotation with non-host crops (e.g. small grains, soy bean or corn) is often recommended to reduce the amount of inoculum in the soil. The following cultural practices are also suggested to prevent and reduce the incidence of black dot.

1. Use certified disease free seed.

2. Treat cut seed with a seed treatment (e.g. Maxim MZ). Although seed treatments may provide limited control of black dot, they improve plant stand and crop vigor, reducing plant stress which increases susceptibility to black dot.

3. Avoid planting in poorly draining soil if possible.

4. Use good crop production practices, such as timely irrigation and adequate fertilization to reduce crop stress.

5. Use tillage practices such as fall plowing that bury plant refuse and encourage decomposition.

6. Harvest tubers as soon as possible after vine kill.

7. Control temperature and humidity in storage. High temperatures and condensation on the tuber surface promotes disease.

Host Resistance

Currently there are no known commercial varieties with resistance to black dot. However, research has shown that in general late maturing varieties tend to be more vulnerable to yield reductions than early maturing varieties. This may be due to the fact that black dot is a late season disease, and leaving tubers in the ground longer exposes them to more disease pressure.

Chemical Control

In furrow applications of azoxystrobin have been reported to reduce or suppress symptoms of black dot on the stem although under conditions conducive for development of black dot, variable results have been reported. Fludioxonil, pyraclostrobin and PCNB have also been reported to suppress black dot.

Beet Vascular Necrosis

Beet vascular necrosis is a disease caused by a bacterium, Erwinia carotovora subsp. betavasculorum, present in many native and cultivated soils.

This pathogen can survive in some weedy hosts. Portions of the bacterium that causes potato blackleg can be pathogenic to sugar beets. Plant wounding, excessive nitrogen or moisture, and warm temperatures (optimum is 79° F to 82° F) favor disease development. The disease occasionally is severe in Idaho.

Symptoms:

Black streaks may be found on petioles, and crowns may be blackened or produce froth. Vascular bundles are brown, and adjacent tissue turns pink when cut and exposed to air. Rot can become extensive soft or dry rot.

Management:

Since the bacteria cannot survive in seeds, the best way to prevent the disease is to ensure that vegetatively propagated plant material are clean of infection, such that the bacterium does not enter the soil. However, if the bacteria is already present, there are some methods that can be used to lessen the infection.

Cultural Practices

Because the bacteria readily enter the plant through wounds, management practices that decrease injury to the plants are important to control the spread of the disease. Cultivation is not recommended, as the machinery can become contaminated and physically spread the bacteria around the soil. Accidental leaf tearing or root scarring can also occur depending on the size of the crop, allowing the bacteria to enter more individual plants. If hilling the beets, great care must be taken to avoid getting soil into the crown, because the pathogen is soil-borne and this could expose the plant to more bacteria, thus increasing the risk of infection.

While most bacteria are motile and can swim, they cannot move very far due to their small size. However, they can be carried along by water, and a significant movement of *Pectobacterium* can be attributed to being carried downstream from irrigation and rainwater. To control the spread of the disease, limiting irrigation is another strategy. The bacteria also flourishes in wet conditions, so limiting excess water can control both the spread and severity of the disease.

Increased in-row spacing also causes more severe disease. In an infected field, yield decreased linearly when spacing was greater than 15 cm (6 in), so a spacing of 6 inches or less is recommended.

The bacteria can also utilize nitrogen fertilizer to accelerate their growth, thus limiting or eliminating the amount of nitrogen fertilizer applied will lessen the disease severity. For example, when fertilizer was applied to an infected field the infection rate per root increased from 11% (with no added nitrogen) to 36% (with 336 kg nitrogen/hectare), and sugar yields decreased.

Resistance

The bacteria can survive in the rhizosphere of other crops such as tomato, carrots, sweet potato, radish, and squash as well as weed plants like lupin and pigweed, so it is very hard to get rid of it completely. When it is known that the bacterium is present in the soil, planting resistant varieties can be the best defense against the disease. Many available beet cultivars are resistant to *Pectobacterium carotovorum* subsp. *betavasculorum*, and some examples are provided in the corresponding table. A comprehensive list is maintained by the USDA on the Germplasm Resources Information Network. Even though some genes associated with root defense response have been identified, the specific mechanism of resistance is unknown, and it is currently being researched.

Cultivar	Resistance
H9	No
H10	No
C17	No
546 H3	Moderate
C13	No
E540	No
E538	No
E534	Moderate
E502	Moderate
E506	Yes
E536	Yes
C930-35	Moderate
C927-4	Moderate
C930-19	Yes
C929-62	Yes

Biological Control

Some bacteriophages, viruses that infect bacteria, have been used as effective controls of bacterial diseases in laboratory experiments. This relatively new technology is a promising control method that is currently being researched. Bacteriophages are extremely host-specific, which makes them environmentally sound as they will not destroy other, beneficial soil microorganisms. Some bacteriophages identified as effective controls of *Pectobacterium carotovorum* subsp. *betavasculorum* are the strains ΦEcc2 ΦEcc3 ΦEcc9 ΦEcc14. When mixed with a fertilizer and applied to inoculated calla lily bulbs in a greenhouse, they reduced diseased tissue by 40 to 70%.ΦEcc3 appeared to be the most effective, reducing the percent of diseased plants from 30 to 5% in one trial, to 50 to 15% in a second trial. They have also been used successfully to reduce rotting in lettuce caused by *Pectobacterium carotovorum* subsp. *carotovorum*, a different bacterial species closely related to the one that causes beet vascular necrosis.

While it is more difficult to apply bacteriophages in a field setting, it is not impossible, and laboratory and greenhouse trials are showing bacteriophages to potentially be a very effective control mechanism. However, there are a few obstacles to surmount before field trials can begin. A large problem is that they are damaged by UV light, so applying the phage mixture during the evening will help promote its viability. Also, providing the phages with susceptible non-pathogenic bacteria to replicate with can ensure there is adequate persistence until the bacteriophages can spread to the targeted bacteria. The bacteriophages are unable to kill all the bacteria, because they need a dense population of bacteria in order to effectively infect and spread, so while the phages were able to decrease the number of diseased plants by up to 35%, around 2,000 Colony Forming Units per milliliter (an estimate of living bacteria cells) were able to survive the treatment. Lastly, the use of these bacteriophages places strong selection on the host bacteria, which causes a high probability of developing resistance to the attacking bacteriophage. Thus it is recommended that multiple strains of the bacteriophage be used in each application so the bacteria do not have a chance to develop resistance to any one strain.

Importance

The disease was first identified in the western states of, California, Washington, Texas, Arizona and Idaho in the 1970s and initially led to substantial yield losses in those areas. *Erwinia caratovara* subsp *betavascularum* was not discovered in Montana until 1998. When it first appeared, beet vascular necrosis caused individual farm yield loss ranging from 5–70% in Montana's Bighorn Valley. Today, yield losses from the disease are generally infrequent and patchy as most producers plant resistant varieties. Infection rate is generally low if resistant cultivars are chosen; however, warmer and wetter conditions can lead to higher than normal instance of disease.

If infection does occur, bacterial root rots can not only cause economic losses in the field, but also can in storage and processing as well. In processing plants, rotten roots complicate slicing and the bacterially-produced slime can clog filters. This is especially problematic with late-infected beets which are generally harvested and processed along with healthy beets. The disease can also lower sugar-content which greatly reduces the quality.

Alternaria Solani

Alternaria solani causes diseases on foliage (early blight), basal stems of seedlings (collar rot), stems of adult plants (stem lesions), and fruits (fruit rot) of tomato.

Hosts and Symptoms

Alternaria solani infects stems, leaves and fruits of tomato (*Solanum lycopersicum* L.), potato (*S. tuberosum*), eggplant (*S. melongena* L.), bell pepper and hot pepper (*Capsicum* spp.), and other members of the *Solanum* family. Distinguishing symptoms of *A. solani* include leaf spot and defoliation, which are most pronounced in the lower canopy. In some cases, *A. solani* may also cause damping off.

On Tomatoes

On tomato, foliar symptoms of *A. solani* generally occur on the oldest leaves and start as small lesions that are brown to black in color. These leaf spots resemble concentric rings - a distinguishing characteristic of the pathogen - and measure up to 1.3 cm (0.51 inches) in diameter. Both the area around the leaf spot and the entire leaf may become yellow or chlorotic. Under favorable conditions (e.g., warm weather with short or abundant dews), significant defoliation of lower leaves may occur, leading to sunscald of the fruit. As the disease progresses, symptoms may migrate to the plant stem and fruit. Stem lesions are dark, slightly sunken and concentric in shape. Basal girdling and death of seedlings may occur, a symptom known as collar rot. In fruit, *A. solani* invades at the point of attachment to the stem as well as through growth cracks and wounds made by insects, infecting large areas of the fruit. Fruit spots are similar in appearance to those on leaves – brown with dark concentric circles. Mature lesions are typically covered by a black, velvety mass of fungal spores that may be visible under proper light conditions.

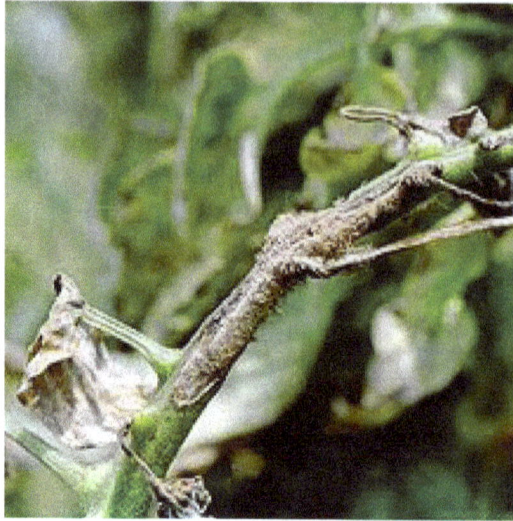
Stem lesion of *Alternaria solani*

On Potatoes

In potato, primary damage by *A. solani* is attributed to premature defoliation of potato plants, which results in tuber yield reduction. Initial infection occurs on older leaves, with concentric dark brown spots developing mainly in the leaf center. The disease progresses during the period of potato vegetation, and infected leaves turn yellow and either dry out or fall off the stem. On stems, spots are gaunt with no clear contours (as compared to leaf spots). Tuber lesions are dry, dark and pressed into the tuber surface, with the underlying flesh turning dry, leathery and brown. During storage, tuber lesions may enlarge and tubers may become shriveled. Disease severity due to *A. solani* is highest when potato plants are injured, under stress or lack proper nutrition. High levels of nitrogen, moderate potassium and low phosphorus in the soil can reduce susceptibility of infection by the pathogen.

Disease Cycle

Alternaria solani is a deuteromycete with a polycyclic life cycle. *Alternaria solani* reproduces asexually by means of conidia. *A.solani* is generally considered to be a necrotrophic pathogen, i.e. it kills the host tissue using cell wall degrading enzymes and toxins and feeds on the dead plant cell material.

The life cycle starts with the fungus overwintering in crop residues or wild members of the Solanaceae family, such as black nightshade. In the spring, conidia are produced. Multicellular conidia are splashed by water or by wind onto an uninfected plant. The conidia infect the plant by entering through small wounds, stomata, or direct penetration. Infections usually start on older leaves close to the ground. The fungus takes time to grow and eventually forms a lesion. From this lesion, more conidia are created and released. These conidia infect other plants or other parts of the same plant within the same growing season. Every part of the plant can be infected and form lesions. This is especially important when fruit or tubers are infected as they can be used to spread the disease.

In general, development of the pathogen can be aggravated by an increase in inoculum from alternative hosts such as weeds or other solanaceous species. Disease severity and prevalence are highest when plants are mature.

Environment

Alternaria solani spores are universally present in fields where host plants have been grown.

Free water is required for Alternaria spores to germinate; spores will be unable to infect a perfectly dry leaf. Alternaria spores germinate within 2 hours over a wide range of temperatures but at 26.6-29.4°C (80-85° F) may only take 1/2 hour. Another 3 to 12 hours are required for the fungus to penetrate the plant depending on temperature. After penetration, lesions may form within 2–3 days or the infection can remain dormant awaiting proper conditions (15.5°C (60° F) and extended periods of wetness). Alternaria sporulates best at about 26.6°C (80° F) when abundant moisture (as provided by rain, mist, fog, dew, irrigation) is present. Infections are most prevalent on poorly nourished or otherwise stressed plants.

Management

Cultural Control

- Clear infected debris from field to reduce inoculum for the next year.

- Water plants in the morning so plants are wet for the shortest amount of time.

- Use a drip irrigation system to minimize leaf wetness which provides optimal conditions for fungal growth.

- Use mulch so spores in soil cannot splash onto leaves from the soil.

- Rotate to a non-Solanaceous crop for at least three years.

- If possible control wild population of *Solanaceae*. This will decrease the amount of inoculum to infect your plants.

- Closely monitor field, especially in warm damp weather when it grows fastest, to reduce loss of crop and spray fungicide in time.

- Plant resistant cultivars.

- Increase air circulation in rows. Damp conditions allow for optimal growth of *A. Solani* and the disease spreads more rapidly. This can be achieved by planting farther apart or by trimming leaves.

Chemical Control

There are numerous fungicides on the market for controlling early blight. Some of the fungicides on the market are (azoxystrobin), pyraclostrobin, Bacillus subtilis, chlorothalonil, copper products, hydrogen dioxide (Hydroperoxyl), mancozeb, potassium bicarbonate, and ziram. Specific spraying regiments are found on the label. Labels for these products should be read carefully before applying.

Quinone outside inhibitor (QoIs) fungicides e.g. azoxystrobin are used due to their broad-spectrum activity. However, decreased fungicide sensitivity has been observed in *A. solani*due to a F129L (Phenylalanine (F) changed to Leucine at position 129) amino acid substitution.

Economic Significance

Early blight caused by *A. solani* is the most destructive disease of tomatoes in the tropical and subtropical regions. Each 1% increase in intensity can reduce yield by 1.36%, and complete crop failure can occur when the disease is most severe. Yield losses of up to 79% have been reported in the U.S., of which 20-40% is due to seedling losses (i.e., collar rot) in the field.

A. solani is also one of the most important foliar pathogens of potato. In the U.S., yield loss estimates attributed to foliar damage, which results in decreased tuber quality and yield reduction, can reach 20-30%.In storage, *A. solani* can cause dry rot of tubers and may also reduce storage length, which both of which diminish the quantity and quality of marketable tubers.

Because *A. solani* is one of numerous tomato/potato pathogens that are typically controlled with the same products, accurately estimating both the total economic loss and the total expenditure on fungicides for control of early blight is difficult. Best estimates suggest that total annual global expenditures on fungicide control of *A. solani* is approximately $77 million: $32 million for tomatoes and $45 million for potatoes.

Tobamovirus

Tobamovirus is a genus in the family Virgaviridae. Tobacco, tomato, potato and squash are natural hosts to this virus. Currently, there are 37 species that are known to belong to this genus. Necrotic lesions on leaves are manifestation of tobamovirus infections on such plants. Tobamoviruses that affect the cucurbits, solanaceous, brassicas and malvaceous plants are four informal subgroups in this genus. These groups differ in terms of their genome sequences and range of host plants.

These viruses are thought to have codiverged with their hosts from a common ancestor. There are at least 3 distinct clades of tobamoviruses, which to some extent follow their host ranges: that is, there is one infecting solanaceous species; a second infecting cucurbits and legumes and a third infecting the crucifers.

Genome

The RNA genome encodes at least four polypeptides: these are the non-structural protein and the read-through product which are involved in virus replication (RNA-dependent RNA polymerase, RdRp); the movement protein (MP) which is necessary for the virus to move between cells and the coat protein (CP). The read-through portion of the RdRp may be expressed as a separate protein in TMV. The virus is able to replicate without the movement or coat proteins, but the other two are essential. The non-structural protein has domains suggesting it is involved in RNA capping and the read-through product has a motif for an RNA polymerase. The movement proteins are made very early in the infection cycle and localized to the plasmodesmata, they are probably involved in host specificity as they are believed to interact with some host cell factors.

Structure

Tobamoviruses are non-enveloped, with helical rod geometries, and helical symmetry. The diameter is around 18 nm, with a length of 300-310 nm. Genomes are linear and non-segmented, around 6.3-6.5kb in length.

Genus	Structure	Symmetry	Capsid	Genomic arrangement	Genomic segmentation
Tobamovirus	Rod-shaped	Helical	Non-enveloped	Linear	Non-Segmented

Life Cycle

Viral replication is cytoplasmic. Entry into the host cell is achieved by penetration into the host cell. Replication follows the positive stranded RNA virus replication model. Positive stranded RNA virus transcription is the method of transcription. Translation takes place by suppression of termination. The virus exits the host cell by monopartite non-tubule guided viral movement. Plants serve as the natural host. Transmission routes are mechanical.

Genus	Host details	Tissue tropism	Entry details	Release details	Replication site	Assembly site	Transmission
Tobamovirus	Plants	None	Unknown	Viral movement	Cytoplasm	Cytoplasm	Mechanical

Routes of Infection

The infection is localized to begin with but if the virus remains unchallenged it will spread via the vascular system into a systemic infection. The exact mechanism the virus uses to move throughout the plant is unknown but the interaction of pectin methylesterase, a cellular enzyme important for cell wall metabolism and plant development, with the movement protein has been implicated.

Leveillula Taurica

Powdery mildew is a serious fungal threat to agricultural production. The endoparasitic powdery mildew fungus *Leveillula taurica* (Lév.) G. Arnaud (anamorph: *Oidiopsis taurica* (Lév.) E. S. Salmon) is an important pathogen of pepper, tomato, eggplant, onion, cotton, and other crops, and it has also been recorded on many wild plant species. This pathogen represents a challenge from many perspectives. First, the early stages of infection are difficult to diagnose; thus, the disease can rapidly spread in both field and greenhouse crop production. Second, the species delimitation in the genus *Leveillula* is problematic and the binomial "*L. taurica*" clearly refers to a species complex that includes several biological species. Moreover, the exact host ranges of the different *L. taurica* lineages recognized by phylogenetic studies are still not known.

Most powdery mildew species are epiparasitic because all their structures except haustoria are developed on the host plant surfaces. In contrast, *Leveillula* and the other genera in tribe *Phyllactinieae*, namely *Phyllactinia* and *Pleochaeta*, develop a partly endophytic mycelium and are endotrophic because their haustoria are produced in the mesophyll cells. Kunoh illustrated that, in *L. taurica*, the germinated conidia are attached to the leaf surface by "adhesion bodies" that differ from appressoria produced by epiphytic species because they do not initiate infection hyphae that penetrate the epidermal cells. Instead, infection hyphae of *L. taurica* enter the host plant through stomata and develop an intercellular mycelium in the spongy and palisade parenchyma tissue with haustoria in some of its cells. Later, conidiophores emerge through stomata, mainly on the abaxial leaf surface, producing primary and numerous secondary conidia that differ in their morphology. At that stage, hyphae are also produced mainly on the abaxial leaf surfaces. On the adaxial surface of infected leaves, chlorotic spots are usually visible, indicating the development of mildew colonies underneath the spots.

Hosts and Symptoms

L. taurica is the pathogen responsible for powdery mildew on onions, but it can also infect peppers, tomatoes, eggplant, cotton, and garlic. While *L. taurica* can infect many different plants it is actually very host specific. Different races of *L. taurica* can only infect certain crops, and even specific cultivars within the same crop. An accurate way to describe its host specificity is that this disease is, "a composite species consisting of many host-specific races." Symptoms of Onion Powdery Mildew (OPM) are usually seen as circular or oblong lesions that are 5 to 20 mm and have a chlorotic or necrotic appearance. The lesions appear on older leaves before the bulb of the onion begins to form, but also can occur on the younger leaves towards the end of the season. As the disease progresses signs of OPM can also be seen. On the lesions white mycelium can be found with conidiophores bearing either lanceolate or rounded condia.

Disease Cycle

The polycyclic disease cycle of *L. taurica* is similar to that of other powdery mildew species. It overwinters (as chasmothecia) in crop residues above the soil surface. Under favorable climatic conditions, the chasmothecia open and release ascospores, which are wind-dispersed. The ascospores enter the host through its stomata, germinate, and colonize the host's tissues with its mycelia. The pathogen then begins to produce its asexual conidia, either singly or on branched conidiophores. The conidia exit through the host's stomata and serve as a secondary inoculum to spread disease after initial infection. In the fall, the pathogen undergoes sexual reproduction and again produces chasmothecia, its dormant, overwintering structure.

Environment

The genus *Leveillula* is distributed in warm, arid areas of Africa, Asia, South America, southern Europe, and the western parts of North America. Species within the genus are adapted to xerophytic conditions, exemplified by the ability of their conidia to germinate rapidly and at any relative humidity. *L. taurica* is primarily a disease of allium species—it has been documented on onions and garlic in Israel and southeastern Europe—but can also infect other species, including cucumbers, peppers, eggplants and tomatoes. It was first reported in the western United States in 1985, infecting onions in the state of California. It has since appeared in Idaho, the state of Washington, and Utah.

Management

OPM tends to appear near the end of the growing season. The best way to control *L. taurica* is to remove all crop residue from the previous onion crop before subsequent planting. Two methods to accomplish this include deep tillage, and rotating to a non-host crop the year following an onion crop. Controlling volunteer onion sprouting (or the emergence of the previous year's onion plants) will also assist in prevention of the pathogen from carrying-over from one year to the next.

Irrigation practices can also be used to limit the development of OPM. Moisture stress has been noted to increase the susceptibility of host species to *L. taurica*. Onions with adequate moisture will be more resistant to the pathogen, and onion crops with overhead irrigation rarely see powdery mildew development in the field.

The fungicide Cabrio (produced by BASF Chemical) is labeled for the control of *L. taurica* on onions, but the disease rarely progresses enough to justify the use of a fungicide. Considerations of economic benefit should be made before the fungicide is applied, and all labeling directions followed.

Resistant varieties have been found in some studies, Jahn et al. found powdery mildew resistance to be extremely beneficial in cucurbits, reducing the need for fungicide, and reducing agricultural losses due to powdery mildew pathogens. Although a truly resistant variety has not been found for onion plants, some onion genotypes with glossy leaves had selective susceptibility to *L. taurica*. Onions with the glossiest leaves were found to be most susceptible, while onions with less glossy leaves showed limited susceptibility. However, the study was unable to come to a conclusion on which variety was best suited for *L. taurica* resistance.

Importance

The economic importance of OPM is limited, as the disease is sporadic, and it rarely progresses enough to make fungicide treatment necessary. Because of the limited importance of OPM, data on incidence rates are not well documented. Simple cultural controls, as mentioned above, are usually effective in controlling losses associated with the disease. The disease geography within the United States is limited to Idaho, Utah, California, and the Pacific Northwest. Findings have also occurred in Israel, Italy, Iran, Sudan, Brazil, and Southeastern Europe.

Didymella Pinodes

Didymella pinodes (syn. *Mycosphaerella pinodes*) is a hemibiotrophic fungal plant pathogen and the causal agent of ascochyta blight on pea. It is infective on several species such as *Lathyrus sativus*, *Lupinus albus*, *Medicago spp.*, *Trifolium spp.*, *Vicia sativa*, and *Vicia articulata*, and is thus defined as broadrange pathogen.

Symptoms

Symptoms include lesions on leaves, stem and pods of plants. The disease is difficult to distinguish from blight caused by *Ascochyta pisi*, though *D. pinodes* is the more aggressive of the two pathogens.

Epidemiology

The disease cycle starts with dissemination of ascospores after which germination pycnidia rapidly develop. Pycnidiaspores quickly disperse by rain splashes are responsible for reinfection over short distances. Consequently, production of pseudothecia is initiated on senescent tissues. After rainfall, ascospores are released from the pseudothecia and disperse by wind over long distances.

Disease Management

Useful levels of resistance remain to be determined and the application of fungicidal sprays was reported to be uneconomical. Furthermore, reports showed that insensitivity arises against chemicals such as strobilurons after continuous application. Thus, cultural management is the preliminar option to control the disease progress by minimizing inoculum carry over as well as survival of inoculum on crop residues and in soil, and avoiding initial infection from arial inoculum. Furthermore, burying of infected residues declines pathogen survival, however, crop rotation and tillage regimes have little influence on disease severity. Delayed sowing by 3–4 weeks reduces ascochyta blight severity by more than 50%, however, such measures are not feasible at higher latitudes, because of a shorter growing season.

Host Resistance

So far, only incomplete resistance is available in the pea germplasm and quantitative differences are highly influenced by environmental conditions, plant age and physiological characteristics of plants. Tall cultivars with more erect growth suffer lower *D. pinodes* infection. Susceptibility increases with earliness and along with maturity of plants.

Besides morphological traits, a proteomic and metabolomic study pinpointed molecular markers contributing to resistance. Disease severity of leaves was also reported to be lower when pea plants are associated with rhizobial bacteria that presumalby provoke so called induced sysmteic resistance.

Ralstonia Solanacearum

Ralstonia solanacearum (Smith) (formerly called *Pseudomonas solanacearum*), is a soilborne bacterial pathogen that is a major limiting factor in the production of many crop plants around the world. This organism is the causal agent of brown rot of potato, bacterial wilt or southern wilt of tomato, tobacco, eggplant, and some ornamentals, and Moko disease of banana.

Symptoms

Above-ground symptoms include wilting of 1-2 leaves on young plants during the heat of the day. Such plants tend to recover at night. On large-leafed plants, only the tissue on one side of the mid-vein may wilt. This is very characteristic for plants such as Nicotiana. Affected leaves turn yellow and remain wilted after a time. The area between leaf veins dies and browns. Usually the main stem of the affected plants remains upright even though all the leaves may wilt and die.

Internal symptoms include light tan to yellow-brown discoloration of the vascular tissue. Long sections of infected stems reveals dark brown to black streaking in the vascular tissue as the disease progresses. As invasion proceeds, the pith and cortex of the stem become dark brown.

Symptoms in geraniums are very similar to those caused by the bacterial blight pathogen, *Xanthomonas campestris* pv. pelargonii (*Xcp*) However, while *Xcp* can cause leaf spotting, *Ralstonia* does not.

Signs of the Pathogen

Slimy, sticky ooze forms tan-white to brownish beads where the vascular tissue is cut. When an infected stem is cut across and the cut ends held together for a few seconds, a thin thread of ooze can be seen as the cut ends are slowly separated. If one of the cut ends is suspended in a clear container clean water, bacterial ooze will form a thread in the water.

Management

Growing and propagating from pathogen-free plant material is the main way to avoid problems with *Ralstonia*, regardless of the race and biovar involved. Propagators must use pathogen-free potting soil or other media, establish stock plants that are tested and known to be free of the bacteria, train workers handling the stock plants in methods and procedures that prevent the pathogen from contaminating the potting soil or coming in contact with the stock plants, and then maintaining this system throughout the propagation phase of crop production.

There are no chemicals or biological agents that adequately control these bacteria. Infected plants must be discarded as soon as possible.

As is the case with all pathogens carried on vegetatively propagated crops, the purchaser of cuttings or pre-finished plants must isolate all new, incoming plants as if the health of the plants were unknown, even if the plants have been certified as healthy. New plants must not be commingled or dispersed among other plants in the greenhouse from other sources. This procedure is crucial because by keeping plants originating from one source together allows you to observe those plants as a group, detect any abnormalities within that group, and treat or discard those plants as a group without affecting or damaging plants from other sources. Keeping them together as a group in a defined area of the greenhouse also limits the area that may need to be quarantined, sanitized, or isolated should a pathogen requiring a 'stop sale' (such as *Ralstonia solanacearum*) be found.

Meloidogyne Enterolobii

Meloidogyne enterolobii was originally described from a population collected from the pacara ear-pod tree (*Enterolobium contortisiliquum* (Vell.) Morong) in China in 1983. In 2001 it was reported for the first time in the continental USA in Florida. *M. enterolobii* is now considered as one of the most important root-knot nematode species because of its ability of reproducing on root-knot nematode-resistant (Mi-1 gene carrying genotypes) bell pepper and other economically important crops.

Management

The most efficient control method is preplant soil fumigation with methyl bromide (Mbr). That can reduce the *M. incognita* reproduction by almost 100%. However, the soil fumigant methyl bromide has been phased out in 2005 because of its negative effects on the ozone layer. A 1995 economic study declared that banning methyl bromide without an alternative method of controlling nematodes would cost the nation's bell pepper industry $127 million in losses.

Some Mbr alternatives have been tested, such as Metham sodium plus chloropicrin (Mna+Pic) and 1,3-Dichloropropene (1,3-D) plus Pic. Mna+Pic provided equal or better *Meloidogyne* control than methyl bromide plus pic, for sting nematode, they are equal to MBR plus pic. Other alternative such as Multiguard, which is a formulation of furfural, a compound derived from sugarcane waste, which has been reported to have both nematicidal and antifungal properties.

Nematode-resistant bell pepper cultivar is another method to control nematode population. Two bell pepper cultivars, *Carolina Wonder* and *Charleston Belle*, have been widely planted in the United States. However, while these varieties offer resistance to *M. incognita*, they are susceptible to *M. enterolobii*.

Crop rotation can be used to control *M. enterolobii*. The root-knot resistant bell peppers are not suggested to be planted in the field all over the seasons because that will select more *M. enterolobii*, which will survive and become a big population. Meanwhile, less severe yield loss of susceptible bell peppers has been observed when growing them after resistant bell pepper.

Alternaria Dauci

A. dauci causes leaf spot and blight in carots. It is present in all carrot production areas of the world, and is capable of rapidly causing severe foliar epidemics.

Hosts and Symptoms

Alternaria Leaf Blight is a foliar disease of carrots caused by the fungus *Alternaria dauci*. *Alternaria dauci* is included in the porri species group of *Alternaria*, which is classified for having large conidium and a long, slender filiform beak. Because many of the members of this group have similar morphology, *Alternaria dauci* has also been classified as formae specialis of carrots, or

A. porri f. sp. dauci. It has been well established that the host range of this disease is on cultivated and wild carrot, but it has also been claimed that *Alternaria dauci* has the ability to infect wild parsnip, celery, and parsley. A study in 2011 by Boedo et al. evaluated the host range of *Alternaria dauci* in a controlled environment and concluded that several non-carrot species could constitute alternate hosts, such as *Ridolfia segetum* (corn parsley) and *Caucalis tenet* (hedge parsley). Despite their findings, reports of *A. dauci* colonization on non-carrot hosts continues to be debated because the use of Koch's Postulates on recovered isolates of *A. dauci* is challenging and is rarely reported; in addition, few reports are often made of such infections in field settings.

Symptoms of *A. dauci* appear first as greenish-brown, then water-soaked, and finally necrotic lesions 8–10 days following an infection event. These lesions will appear on carrot leaflets and petioles, and have a characteristic chlorotic, yellow halo. The lesions can be irregularly shaped, and will often appear on older leaves first. Older leaves are the most susceptible to infection; when approximately 40% of the leaf surface area has become infected by *Alternaria dauci*, the leaf will completely yellow, collapse, and die. It is during extended conditions of warm, moist weather that lesions can coalesce and cause entire tops of carrot plants to die off, a phenomenon that is sometimes mistaken for frost damage. The symptoms of this disease are also commonly confused with Cercospora Leaf Blight of carrots as well as bacterial blight, and microscopic analysis is frequently needed to accurately diagnose the pathogen. *A. dauci* produces characteristically dark to olive-brown hyphae and elongated conidiophores, with conidia typically borne singly. Petiole infection can also occur without any lesion development on leaflets, and *A. dauci* can additionally result in damping-off of seedlings, seed stalk blight, and inflorescence infections. These symptoms can significantly reduce yield due to lost photosynthetic activity, prevention of mechanical harvest, and infection of commercial carrot seeds.

Disease Cycle

Alternaria dauci lesion

Sexual reproduction of *Alternaria dauci* is not known to occur, and the disease is most active during spring, summer, and autumn cropping cycles. The disease cycle begins when fungus overwinters on or in host seed and in soil-borne debris from carrot. *A. dauci* may also be spread into fields via contaminated carrot seeds during cultivation. Once introduced, the pathogen can persist in carrot debris or contaminated seeds in the soil for up to two years. Seedling infection near the hypocotyl-root junction (just below the soil line) then occurs in the early spring following over-wintering of *Alternaria dauci* mycelium or conidia. This infected region will become necrotic and lead to the production of more asexual conidia on conidiophores, which will serve as secondary inoculum. Wind and rain cause conidia to disperse to neighboring host species, and multiple germination tubes will be produced from each conidium that successfully colonizes a new host. As penetration occurs, *Alternaria dauci* will produce a chemical known as phytotoxin zinniol, which degrades cell membranes and chloroplasts, ultimately leading to the chlorotic symptoms characteristic of the disease. These germination tubes will pierce host cell walls to initiate infection, or if wounds are present the pathogen may enter in that manner. The process of germination, penetration, and symptom development generally occurs in a timespan of 8 to 16 days, but the presence of wounds shortens the amount of time needed to carry out the process.

Following these events, conidia are repeatedly produced from leaf and stem lesions throughout the summer months, allowing the pathogen to be dispersed to its surrounding environment. Inflorescence that is infected by *A. dauci* early in the summer will produce nonviable seeds, but plants infected later in the summer or early fall may still carry viable seeds; this fungus remains in the pericarp and does not penetrate the embryo or endosperm (non-systemic). Following harvest in the fall, *Alternaria dauci* will persist in remaining carrot debris in the soil or be concentrated in infected seedlings, and the disease cycle will be repeated.

Environment

Production and transmission of *Alternaria dauci* is heightened during moderate to warm temperatures and extended periods of leaf wetness due to rainfall, dew, or sprinkler irrigation. Infection can occur between temperatures of 57 - 95 degrees Fahrenheit, with 82 degrees Fahrenheit being optimal. Mycelium and spores are spread through splashing rain, tools for cultivation or contaminated soils. *Alternaria* diseases, in general, tend to infect older, senescing tissues, and on plants developing under stress. A study conducted by Vital et al. in 1999 assessed the influence of the rate of soil fertilization on the severity of Alternaria Leaf Blight in carrots and found low levels of nitrogen and potassium increased the severity of the pathogen, while high levels of the nutrients reduced disease severity. Though it is not well understood why this occurs, it is postulated that higher nitrogen levels may extend a plant's vigor and delay maturation, which is important because *A. dauci* is more likely to infect senescing tissue.

Management

Effective management for *Alternaria dauci* involves preventing the introduction and development of the disease. One of the best practices to avoid infection is to plant pathogen-free seed or seed treated with hot water at 50 degrees Celsius for twenty minutes. In addition to seed treated with fungicide or hot water, once harvest is complete it is imperative to turn the carrot residue under the soil. The pathogen only survives on infected plant debris, allowing this practice to hasten decomposition of the debris. Crop rotation will allow the debris enough time to decompose. Recommendations vary depending on location, but 2 years is the minimum allowance for rotation. Planting carrots continuously in the same field will result in increased infection. New fields should not be located near previously infected fields in order to prevent contamination through dispersal. Dispersal can occur through multiple avenues such as rain splash, farm equipment, workers, and insects.

Cultural practices can also promote reduction of *Alternaria dauci*. They include practices that will lower the duration of leaf wetness and soil moisture. Planting on raised beds with wider row spacing has been shown to reduce soil moisture, thereby limiting the spread of the disease. Symptoms tend to be more severe on carrots that are stressed or poorly fertilized. In order to avoid more severe symptoms, keep the plants free of injury, watered, and adequately fertilized. Although resistant varieties are not available, the susceptibility of the carrot differs by variety. The varieties least susceptible vary by state, and a list of varieties appropriate to a specific area can be found through the state's extension program. In the Midwest, the University of Wisconsin and the University of Michigan have bred varieties including Atlantis, Beta III, and Chancellor that exhibit resistance.

In the absence of treated seed, there are multiple chemical sprays available to treat *Alternaria dauci*. Azoxystrobin, chlorothalonil, iprodione, pyraclostrobin and bacillus are a few common fungicides to consider for foliar application. A few brand names to look for in the Midwest include RR Endura 70 WG, Rovral, and Switch. Gibberrillic acid has shown to be equally effective as the aforementioned fungicides. However, if sprayed in excess giberrillic acid can defer nutrients from the roots to foliage, resulting in undeveloped carrots. If chemically treating plants, scouting the crop is of utmost importance. Initial threshold recommendations vary depending on location, time of year, and moisture level. Different recommendations include spraying upon first evidence of symptoms or spraying once disease has reached 25% of the foliage.

Importance

Alternaria dauci is one of two leading pathogens affecting carrots around the world. Most often found in temperate climates, the disease has been found in North America, the Netherlands, the Middle East, and even parts of Southern Asia and India. Carrot leaf blight is especially damaging in that its leaf lesions not only reduce photosynthetic area, but also weaken the leaves and petioles structurally. This makes mechanical harvesting of the carrot crop less efficient, and yields are even worse when blighted leaves have been exposed to heavy frosts.

Alternaria dauci can spread rapidly if not controlled. Between February and November 2003, when the disease first spread to Turkey, 73-85% of surveyed fields were shown to be infected. Of those fields, disease rates among individual plants ranged from 65-90% total infection within the field. The highest levels of occurrence were always in moist fields with low levels of drainage.

Aster Yellows

Aster yellows is a plant disease caused by a phytoplasma bacterium, which affects over 300 species of herbaceous broad-leafed plants. Aster yellows is found over much of the world wherever air temperatures do not persist much above 32°C (90° F). As its name implies, members of the family Asteraceae are vulnerable to infection, though the disease can also affect a variety of common vegetables like, lettuce, carrot, tomato, and celery, cereals, garden plants, and wild species.

Typical symptoms include yellowing (chlorosis) of young shoots, stiff and erect bunchy growth, greenish and distorted or dwarfed flowers, and general stunting or dwarfing. The phytoplasma lives in the phloem of infected plants and is transmitted by leafhopper insects when they feed on an infected plant and then on a healthy one. No transmission occurs through leafhopper eggs or plant seed. The phytoplasma is perpetuated in overwintering weed and crop plants, in propagative parts (bulbs, corms, tubers), and in leafhoppers in mild climates. The phytoplasma is destroyed in plants and leafhoppers subjected to temperatures of 38 to 42°C (100 to 108° F) for two to three weeks; thus, aster yellows is rare or unknown in many tropical regions.

Though the disease is not lethal, control is effected chiefly by promptly removing diseased plants and all overwintering susceptible weeds. Spraying or dusting with a contact insecticide repulses the leafhopper carriers.

Symptoms of Aster Yellows

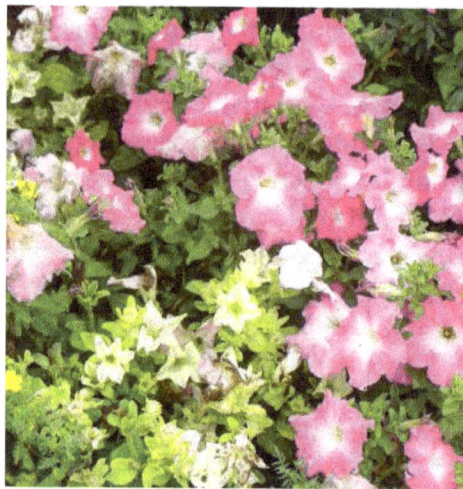

Infected petunia did not develop to normal size or color

- Leaves are discolored pale green to yellow or white.

- In some plants, red to purple discoloration of leaves occurs.

- Leaves may be small and stunted.

- Flowers are small, malformed and often remain green or fail to develop the proper color.

- Plants infected early in the growing season may remain small and stunted.

- Many thin, weak stems grow close together forming a witches' broom.

- Tap roots of carrots are thin, small, covered in many root hairs, and often taste bitter.

Integrated Pest Management Strategies

1. Remove diseased plants: Once a plant is infected with aster yellows, it is a lost cause since the disease is incurable. Early diagnosis and prompt removal of infected plants may help reduce the spread of the disease. Although the disease itself is not fatal to the plant, its presence makes it impossible for a plant to fulfill its intended role in the garden.

2. Plant less susceptible plant species: Controlling aster yellows is difficult. As long as infected leafhoppers are around, they can infect plants. A practical way to avoid having problems with this disease is to grow plants that are not as susceptible to aster yellows. Verbena, salvia, nicotiana, geranium, cockscomb, and impatiens are among the least susceptible plants.

3. Control insects: Vegetable growers may protect susceptible crops by using the mesh fabrics that keep leafhoppers and other insects away from the plants. Some growers put strips of aluminum foil between rows because bright reflections of sunlight confuse the leafhoppers.

4. Control weeds: Remove weeds in your lawn, garden, and surrounding areas, including plantain and dandelion that may harbor the disease.

Phytophthora Capsici

Phytophthora capsici infects more than 50 plant species in more than 15 families. Among the affected plants, cucurbits and peppers are the most susceptible hosts.

Symptoms and Signs

Phytophthora blight, caused by the oomycete plant pathogen *Phytophthora capsici*, can develop on cucurbit plants at any stage of development. The pathogen can infect seedlings, vines, leaves, and fruit. The infection usually appears first in low areas of the fields where soil remains wet longer.

Damping-off: *Phytophthora capsici* causes pre- and post-emergence damping-off in cucurbits under wet and warm [20-30°C (68-86° F)] soil conditions. In seedlings, a watery rot develops on the hypocotyl at or near the soil line, resulting in plant death. Mature plants show symptoms of crown rot. Post-emergence plant death is preceded by plant wilting: a sudden, permanent wilt of the plant without a change in color of the foliage. Leaf wilting progresses from the base to the extremities of the vines. Plants often die within a few days of the first symptoms expression or after soil is saturated by excessive rain or irrigation. The stems of infected plants turn light to dark brown near the soil line and become soft and water-soaked. Infected stems collapse and die. The taproot and lateral roots of infected processing pumpkin plants usually do not exhibit symptoms. Following death of the foliage, roots may give rise to new vines if environmental conditions become less conducive for disease development. Phytophthora damping-off may result in partial to total loss of the crop.

Vine blight: Vines can be affected at any time during the growing season. Water-soaked lesions develop on vines. The lesions are dark olive and then become dark brown in a few days. Lesions girdle the stem, resulting in rapid collapse and death of foliage above the lesion.

Leaf symptoms: *Phytophthora capsici* can infect both the petioles and the leaf blades of plants. Dark brown, water-soaked lesions develop on petioles (similar to lesions on vines), resulting in rapid collapse of the petiole and leaf death. Infected leaf blades develop spots ranging from 5 mm (0.2 in.) to more than 5 cm (2 in.) in diameter. Infected areas are chlorotic at first, but within a few days they become necrotic with chlorotic to olive-green borders. Under wet and warm conditions, leaf spots expand rapidly, coalesce, and may cover the entire leaf. Under dry conditions, leaf spots cease to expand.

Fruit rot: Fruit rot can occur at any time from fruit set until harvest. Fruit rot generally starts on the site of the fruit that is in contact with the ground. However, occasionally infections will begin in other locations on the fruit where infected leaves or vines come into contact with a fruit. Also, symptoms on the upper surface of the fruit develop following rain or overhead irrigation, which can splash water containing the pathogen onto neighboring plants. Fruit rot also can develop after harvest, during transit or in storage. Fruit rot typically begins as a water-soaked lesion. Lesions expand, and become covered with white mold. The pathogen produces numerous sporangia on most infected fruit. Fruit infection progresses rapidly, resulting in complete collapse of the fruit. Phytophthora foliar blight and fruit rot may result in total loss of the crop.

Pathogen Biology

Phytophthora capsici is classified in the family Pythiaceae, order Peronosporales, and class Oomycetes. Oomycetes are not true fungi and have been placed in the kingdom Stramenopila. They are more closely related to brown algae than to true fungi. The pathogen produces asexual sporangia and biflagellate zoospores and sexual oospores. Mycelia are coenocytic (non-septate).*Phytophthora capsici* grows at 10 to 36°C (50 to 97° F), with optimal temperatures of 24 to 33°C (75-91° F). This pathogen grows rapidly on lima bean agar, and the colony diameter can reach up to 8 cm (3 in.) in 5 days. The growth patterns of colonies can vary from cottony, petaloid, rosaceous, to stellate (star-shaped).

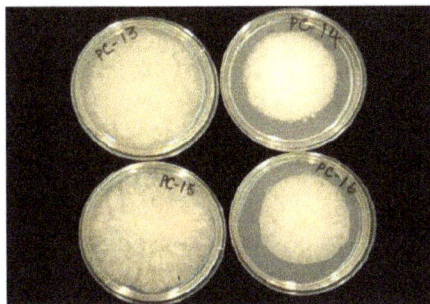

Sporangia (asexual fruiting bodies) of *P. capsici* are produced on sporangiophores (sporangia-producing hyphae) and are mostly papillate (having a small rounded protuberance). Sporangial shapes are influenced by light and other cultural conditions, and may appear as sub-spherical, ovoid, obovoid, ellipsoid, fusiform, or pyriform. The lengths and widths of sporangia can vary from 32.8 to 65.8 and 17.4 to 38.7 µm, respectively. Length/width ratios of sporangia range from 1.3:1 to 2.1:1. Sporangia have long pedicels (stalks), ranging from 35 to 138 µm. Pedicellate sporangia can be dispersed in wind driven rain. Under moist conditions, zoospores (asexual spores) are produced inside sporangia. Zoospores are single-celled and biflagellate.*Phytophthora capsici* also produces chlamydospores (thick-walled asexual spores), which may be terminal or intercalary (between cells) on the mycelium. Chlamydospores can range in diameter from 22 to 39 µm.

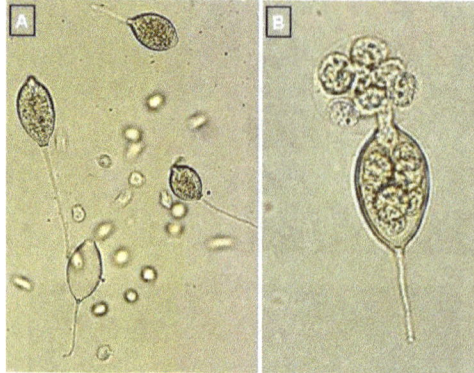

Phytophthora capsici produces sexual structures called antheridia and oogonia, and sexual spores called oospores. *Phytophthora capsici* is predominantly heterothallic with two mating types known as A1 and A2. Antheridia are amphigynous (forming a collar at the base of the oogonium after the young oogonium grows through it), with diameters of 12–21 to 12–17 µm. Oogonia are spherical or sub-spherical, with diameters ranging from 23 to 50 µm. Oospores are predominantly plerotic (filling the oogonium) with wall thicknesses ranging from 2 to 6 µm, and diameters ranging from 22 to 35 µm.

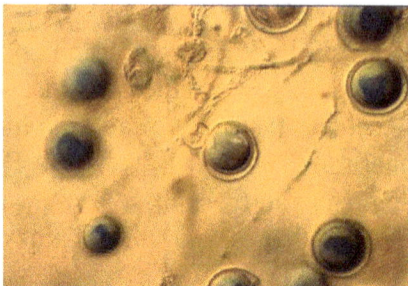

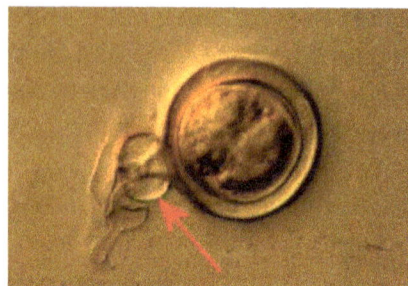

Phytophthora capsici is distinguished from other *Phytophthora* species by its sporangial morphology. Sporangia of *P. capsici* are caducous (easily separated from sporangiophores), have long pedicels, and are spherical to elongate with a tapering base.

Significant differences in virulence (degree of pathogenicity) and genetics among isolates of *P. capsici* have been reported. Several methods can be used to study the genetic variation of *P. capsici* and other fungi. Sequencing and/or restriction digest of internal transcribed spacers (ITS) regions can be used for species identification. A specific PCR primer (Pcap) has been developed that can be used with iTS primers to specifically amplify *P. capsici*. Inter-simple sequence repeats (ISSR) amplification, amplified fragment-length polymorphism (AFLP), allozyme genotyping, and restriction fragment length polymorphisms with a probe can be used to study genetic variation among populations of *P. capsici*.

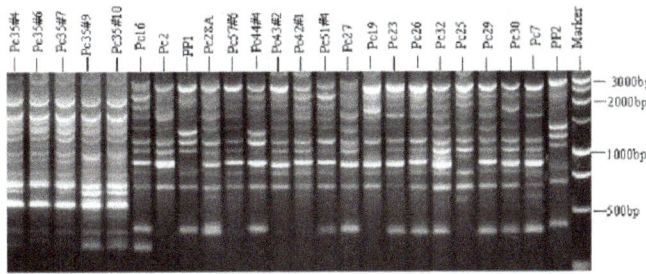

Disease Cycle and Epidemiology

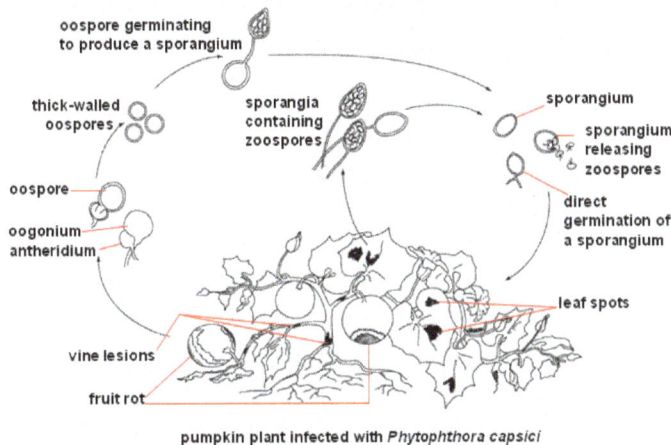

pumpkin plant infected with *Phytophthora capsici*

Phytophthora capsici is a soilborne pathogen and survives between crops as oospores in soil or mycelium in plant debris. Oospores are resistant to desiccation, cold temperatures, and other extreme environmental conditions, and can survive in the soil, in the absence of a host plant, for several years. Oospores germinate and produce sporangia and zoospores. Zoospores are released in water and dispersed by irrigation or surface water. Zoospores are able to swim for several hours and infect plant tissues. Zoospores first lose their flagella and then encyst and form a cell wall, germinate and infect plant tissues. Abundant sporangia are produced on infected tissues, particularly on affected fruit. Sporangia are dispersed by water or in wind-driven rain in the air. Sporangia may either germinate directly and infect the host plant or germinate and give rise to zoospores that are released in water and infect the plant. The pathogen grows within the host and produces sporangia on the surface of the infected tissues. If the environmental conditions are conducive, the disease develops rapidly. Although the pathogen produces chlamydospores on culture media, their role in pathogen survival and diseases epidemiology is not known.

Soil moisture conditions are important for disease development. Sporangia form when soil pores are drained, and they release zoospores when soil is saturated (soil pores are filled with water). The disease is usually associated with heavy rainfall, excessive-irrigation, or poorly drained soil. Frequent irrigation increases the incidence of the disease. Warm conditions are favorable for disease development.

Disease Management

No single method is available to provide adequate control of Phytophthora blight. Various disease control practices can be integrated to manage Phytophthora blight, including: exclusion, cultural practices, and chemical control.

Exclusion

The most effective method of control for Phytophthora blight is to prevent *P. capsici* from moving into a non-infested field.*Phytophthora capsici* spreads by soil, water, and/or plant material. It is highly recommended to thoroughly clean all farm equipment that is used in an infested field before moving it to another field. Also, avoid using water sources (i.e. ponds or reservoirs) that receive run-off water from an infested field. Water sources can be tested for the presence of the pathogen by baiting techniques.*Phytophthora capsici* is not considered a seed-borne pathogen, however, saving seed from a field where Phytophthora blight occurred should be avoided.

Cultural Practices

The following cultural practices can help to manage Phytophthora blight in cucurbit fields. Because *P. capsici* can survive in soil for several years, fields without a history of Phytophthora blight should be selected for planting. Although no cropping rotation period has been established for effective management of Phytophthora blight of cucurbits, it is recommended to select only fields that have not had a history of cucurbits, eggplant, peppers, and/or tomatoes for at least 3 years. Fields should be selected that are well isolated from fields infested with *P. capsici*. High soil moisture favors the development of Phytophthora blight, thus well-drained fields should be selected and excessive irrigation should be avoided. Also avoid planting cucurbit crops in areas of the field that have poor drainage.

Non-vining cucurbit crops (e.g. summer squash) should be planted on dome-shaped raised beds [approximately 25 cm (10 in. high)]. The field should be scouted regularly for Phytophthora symptoms, especially after major rainfalls, and particularly in low areas of the field. When symptoms are localized in a small area of the field, the infected plants should be plowed into the soil. Plants should be sprayed with effective fungicides at the first sign of the disease. Healthy fruit should be removed from the infested area as soon as possible, and they should be checked for disease development routinely. Growing cover crops and/or mulching with plant materials including straw and rye vetch can also be used to manage the dispersal of the pathogen.

Chemical Control

Fungicide seed-treatment and spray-application can prevent seedling death and reduce foliar blight and fruit rot. Seed treatment with either mefenoxam [Apron XL LS at the rate of 0.42 ml / kg (0.64 fl oz/100 lb) seed] or metalaxyl [Allegiance FL at the rate of 0.98 ml /kg (1.5 fl oz/100 lb) seed] can protect seedlings of cucurbits against *P. capsici* for up to 5 weeks after planting. Spray applications of dimethomorph [Acrobat 50WP at the rate of 448 g /ha (6.4 oz/A)] plus copper sulfate [e.g. Cuprofix Disperss 36.9F at the rate of 2.25 kg/ha (2 lb/A)], at weekly intervals, can provide effective protection against foliar blight and fruit rot caused by *P. capsici* in cucurbit fields. Combining Apron XL LS seed-treatment with spray-applications of Acrobat plus copper can minimize crop losses to Phytophthora blight in cucurbit fields. It is important to note that resistance to both mefenoxam and metalaxyl has occurred in some areas of the US, so the sensitivity of *P. capsici* populations should be tested before fungicide applications are chosen.

References

- David J. Hunt & Zafar A. Handoo (2009). "Taxonomy, identification, and principal species". In Roland N. Perry, Maurice Moens & James L. Starr. Root-knot Nematodes. CAB International. pp. 55–97. ISBN 978-1-84593-492-7

- Sugar-beet-beta-vulgaris-bacterial-vascular-necrosis-rot-erwinia-root-rot: pnwhandbooks.org, Retrieved 21 June 2018

- Lunt DH. Genetic tests of ancient asexuality in root knot nematodes reveal recent hybrid origins. BMC Evol Biol. 2008;8:194–216. doi: 10.1186/1471-2148-8-194

- Bacterial-wilt-ralstonia-solanacearum: extension.psu.edu, Retrieved 15 April 2018

- Farrar, James J.; Pryor, Barry M.; Davis, R. M. (August 2004). "Alternaria Diseases of Carrot". Plant Disease. 88: 776–784. doi:10.1094/PDIS.2004.88.8.776. Retrieved Oct 11, 2015

- Carrot-alternaria-leaf-blight, fact-sheets, vegetable: umass.edu, Retrieved 19 June 2018

- Gleason, Mark (May 2000). "Influence of Gibberellic Acid on Carrot Growth and Severity of Alternaria Leaf Blight". APS Journal. 84: 555–558. doi:10.1094/PDIS.2000.84.5.555

- Aster-yellows, science: britannica.com, Retrieved 11 July 2018

- J. N. Sasser & C. C. Carter (1985). An Advanced Treatise on Meloidogyne. Volume I: Biology and Control. North Carolina State University. ISBN 0-931901-01-4

Vegetable Preservation Methods

Vegetables are preserved to extend their availability for consumption. Deterioration in quality occurs as a result of the action of microorganisms or due to naturally occurring enzymes. Some of the different vegetable preservation methods are freezing, canning, pickling and fermenting of vegetables which have been carefully analyzed in this chapter.

Vegetable Preservation

The best food preservation method for fresh vegetables depends on their degree of ripeness. To preserve the best quality vegetables, it helps to understand the difference between maturity and ripeness. Maturity means the produce will ripen and become ready to eat after you pick it. Ripeness occurs when the color, flavor, and texture is fully developed. Once it is fully ripe, fresh produce begins the inevitable and declining spoilage process. Here's a guideline:

- Mature, slightly under ripe produce is optimal for canning and pickling.
- Ripe produce is best for fresh eating, drying, and freezing.
- Overripe produce is suitable for cooking and freezing; cook vegetables into soup or stew.
- Moldy or decaying produce belongs in the composter or worm bin!

To prepare fresh vegetables for preserving, always wash in plenty of running water, remove non-edible parts such as stems and seeds, peel or trim as desired, and cut into slices or cubes

Common Ways to Preserve Vegetables

Canning

There are two primary methods of canning: a hot water bath and pressure canning. Whichever method you use, be sure to use jars with lids made specifically for that technique. Glass canning

jars, which are reusable, come in various sizes (most are single pints or quarts), so choose one that best suits your canning needs. Do not use jars larger than specified in the recipe you follow, as an unsafe product may result.

While most people think of canned foods as salty, all that sodium is optional when you do it yourself. Just make sure that you use "canning salt" not table salt if you plan to salt your foods because regular table salt can make your vegetables soggy. Another tip: Wipe down your rims before you apply the lids and rings as a tight fit is vital for a safe seal.

The hot water bath canning method is for foods that are acidic (pH below 4.6), such fruits, pickles, sauerkraut, jams, jellies, marmalades, and fruit butters. If you are making jams or jellies, it is important that you sterilize the jars, lids, and rings for 10 minutes in boiling hot water before using them. Most fruits and vegetables will last up to 12 months when canned using this method.

Supplies you will need:

- A large pot
- Sterilized jars, lids, and rings
- Thermometer
- Jar rack and/or jar lifter (jar grabbing tongs)
- The foods you are canning

Ways to do it:

1. Begin by following the directions on your preferred recipe for jam, jelly, sauce, canned vegetables, etc. Prepare your fruits and/or vegetables according to the recipe and fill your sterilized jars with the final product, as indicated by the recipe. Add the sterilized lid and ring and tighten.

2. Fill your large pot halfway with water and preheat it to 140-180 degrees Fahrenheit.

3. Add your canned goods (complete with lids) to the pot. Some canning-specific pots come with a rack that you can load the jars into, which makes for easy removal of the hot jars. If you don't have such a rack, simply place the jars one by one into the water (and later remove with a jar-lifting set of tongs).

4. Add boiling water to the pot to bring the water level to 1 inch above the submerged jars; bring the whole pot to a vigorous boil.

5. As soon as the water begins to boil, start the timer. Cover and reduce the heat to maintain a low boil and process for the recommended time (according to your recipe).

6. When the time is up, carefully remove the jars to cool on a towel or cooling rack. Use extreme caution, as the contents will be very hot! If you have done it correctly, the lids should be sealed and concave. Check the seals after 12-24 hours.

The pressure canning method is necessary for any foods that are low acid (pH greater than 4.6) because these foods are not acidic enough to prohibit the growth of bacteria (such as Clostridium botulinum, which grows into botulism and causes extreme and potentially fatal food poisoning). Low-acidic foods include red meats, seafood, poultry, milk, and all fresh vegetables except most tomatoes. In addition, all foods that can be canned with the hot water bath method (above) can also be processed using this method.

The heat, up to 240 degrees Fahrenheit, and pressure generated by using the pressure canning method should be effective in killing all harmful bacteria. It isn't necessary to sterilize the jars, lids, and rings when using this method as the canning process itself will kill all harmful bacteria.

Pressure canning prevents most foods from spoiling altogether, extending their shelf life longer than many other preserving techniques do. However, you will need to invest in a pressure canner. These can be expensive, but when well cared for, they will last for generations. Most are made of aluminum or stainless steel and come with a locking lid that is vented for steam, a jar rack, an automatic vent, a pressure gauge on top and a safety fuse. Make sure you have read the instructions that accompany your pressure canner so that you fully understand how to use it before attempting to do so!

Supplies you will need:

- Pressure canner

- Jars, lids, and rings

- Jar lifters

- The foods you are canning

Ways to do it: Follow the directions in your manual to determine how many cups of water to add to your pot before you start. Unlike the hot water bath method, pressure canning does not require jars to be fully submerged in water—usually just 2-3 cups.

1. Place the jar rack down into the water and, using your jar lifters, place the filled jars down into it.

2. Fasten the lid securely and vent it according to your manual.

3. Heat the water to a boil until steam flows out, then leave the weight off the vent port (or petcock depending on your pressure canner). At this point, you will probably hear a hissing noise.

4. Turn your burner up as high as it will go until steam starts coming out of the vent (or pet-cock) for 10 straight minutes (or as directed in your manual).

5. Next, pressurize your canner. Close the petcock or put the weight on and watch the gauge begin to rise to your desired pressure. Once it reaches that pressure, start timing (duration varies by jar size, contents and altitude, but it is often between 5 and 15 minutes). Adjust your burner as needed to maintain the pressure.

6. Once finished, turn off the burner and allow the pressure to normalize before removing lid. Use extreme caution when removing the jars; the steam can burn and the contents of the jars will be very hot! Place jars onto a towel or cooling rack.

Freezing

Freezing is a good option for fruits you like adding to smoothies or baked goods (bananas, berries, cherries, etc.) and those that aren't suitable for canning. Vegetables such as broccoli, beans, carrots, peas, and corn freeze well, too. Freezing is quick and requires little in the way of equipment or skill, but frozen foods don't last as long as canned foods. Plus, some integrity is lost (foods darken or develop a mushy texture) after freezing.

Supplies you will need:

- Flat baking sheets (or similar containers) that fit into your freezer
- Freezer bags or reusable containers that have tight-fitting lids
- Permanent marker and labeling supplies
- The foods you are freezing

Ways to do it: Many vegetables will require a short blanching (a short boil) before freezing. Beyond that, the method of freezing and storing vegetables is the same as that of fruit (below).

1. Wash, core, and skin (if needed) your fruit. Cut fruit into slices or chunks, if desired.

2. If you are concerned about browning, you can soak the fruit in water with a bit of lemon juice; commercially made agents are available for this purpose, too.

3. Lay prepared fruit on several baking sheets in a single layer. Make sure your fruit is patted dry or unnecessary ice crystals will form.

4. Place baking sheets into the freezer, making sure no fruit is touching, for several hours.

5. Once frozen, remove the fruit and place it into storage bags or containers that are clearly labeled with the contents and the date.

Dehydrating (Drying)

Dehydrating removes all the water from a food and because it lacks moisture, mold and bacteria can't grow on it. Dehydrated foods will last about four months to a year, but some nutrients will be lost in the process. Commonly dried foods include meats, fruits (either in their original form or pureed to make fruit leathers or bars), herbs and seeds.

In hot, arid regions, sun drying is an option, but it demands at least 3-4 sunny days of 100-degree heat in a row. The easiest and most effective way of drying your foods is to use a commercially made dehydrator. These have several levels of stacking trays that allow air to circulate in and around the foods at just the right temperature—high enough to dehydrate the food but low enough not cook it. Generally, the foods are laid out on the trays and, according to the manufacturer's instructions, you'll set the time, temperature and position of the trays. Dehydrators can take several hours or days to dry foods completely. Once dried, keep all foods tightly sealed in a container in a cool, dark place to ensure its longevity.

Some food preservation books and "raw" food cookbooks also include detailed instructions for using a conventional oven, set at a low temperature with the door cracked, as a food dehydrator, which is a great option if you're not ready to invest in your own dehydrator.

Pickling

Pickling, which uses salt and/or vinegar to inhibit the growth of bacteria, is one of the oldest methods of food preservation. While most of us think of sweet pickled cucumbers, sauerkraut, relishes and fruits can also be pickled. Pickled foods will last anywhere from 3 months to a year. There are many recipes and methods for pickling, but most include brining (soaking a food in a salt solution, similar to marinating) for several hours or even days.

Trust the instructions given in pickling recipes, as altering the ratios can be harmful. Do not use table salt; use "canning salt" or "pickling salt" instead. White distilled and cider vinegars of 5 percent acidity (50 grain) are recommended. Another tip: If using cucumbers to make your own pickles, you must remove and discard a 1/16-inch slice from the blossom ends of each cuke. (Blossoms may contain an enzyme that causes excessive softening of pickles.)

Supplies you will need:

- A large pot or pressure canner
- Sterilized jars, lids, rings
- Jar lifters
- Vinegar
- Canning or pickling salt
- Spices (according to recipe)
- The foods you are pickling

Low Cost Technology for Preservation of Vegetables

Vegetables like cabbage, cauliflower, green papaya, bean, pea, carrot, turnip, radish, pumpkin etc, can be preserved by this technique up to a period of about three months without much alteration in the nutritive value of the produce.

Process

- Wash the vegetables thoroughly in clean water. Remove blemishes, rotten part, if any.
- Shred the vegetables, preferably with a stainless steel knife.
- Weigh 2 kg finely (1/8" to 1/16" thick) shredded cabbage and keep it in a clean plastic sheet or container and mix it with 2 kg shredded vegetables (almost any type of vegetables may be used except potato, sweet potato etc.). The proportion of cabbage may be more but not less than 50%. Cabbage contains desirable lactic acid bacteria and the nutrients which help in fermentation. (When cabbages are not available, radishes or cucumbers may serve the same purpose).
- Add 22.5 gm of salt (NaCl) per Kg of shredded vegetables. Mix thoroughly for 3 to 5 minutes. Put the mixture in plastic/earthen/wooden buckets. (Wax lining is necessary for-earthen and wooden vats). Press vegetable mix with hand so that brine can come up at the top of the vegetables.
- Cover the container with a plastic sheet (200 gauge) which touches the surface of vegetables to avoid contact with air.
- Pour water (which must not mix with the vegetables) at the top of plastic sheet so that adequate pressure on the vegetables is ensured. Fasten the bucket with a thread around the neck so that the entire system becomes almost air-light.

Precautions

- Keep the bucket in a safe place. Don't open the cover of the bucket. This will spoil the product. Vegetables can be kept in this way for a period of about three months.
- Once it is opened, the materials should be consumed on the same day. If it is not possible

special presentation techniques like heat processing (bottling), refrigeration or addition of preservatives may be applied to preserve it for longer period of storage for marketing.

- Vegetables get sour by fermentation. If the product is too sour it may be washed thoroughly in water/hot water to make it free from acid and salt. Boil the vegetables and cook with spices to suit individual taste.

Advantages

Fermented juice in the vat is also nutritionally sound. This has a characteristic odour and is rich in vitamins B and C. The liquid has mild laxative effect. Heat the juice for a few minutes after adding a little sugar and spices and use as a soft drink. It is also a good appetizer.

Frozen Vegetables

Freezing food preserves it from the time it is prepared to the time it is eaten. Since early times, farmers, fishermen, and trappers have preserved their game and produce in unheated buildings during the winter season. Freezing food slows down decomposition by turning residual moisture into ice, inhibiting the growth of most bacterial species. In the food commodity industry, the process is called IQF or Individually Quick Frozen. Preserving food in domestic kitchens during the 20th and 21st centuries is achieved using household freezers. Accepted advice to householders was to freeze food on the day of purchase. An initiative by a supermarket group in 2012 promotes advising the freezing of food "as soon as possible up to the product's 'use by' date". The Food Standards Agency was reported to support the change, providing food has been stored correctly up to that time. Frozen vegetables can actually be more nutritious than fresh, since they're packaged immediately after harvesting and the nutrients stay at their peak. Vegetables typically last for about eight months unopened in the freezer.

Health Benefits and Risks

In general, boiling vegetables can cause them to lose vitamins. Thus, the process of blanching does have deleterious effects on some nutrients. In particular, vitamin C and folic acid are susceptible to loss during the commercial process, and canned or frozen broccoli for instance loses the entirety of its most valuable nutrient. In addition, studies have shown that thawing frozen vegetables before cooking can accelerate the loss of vitamin C.

Over the years, there has been controversy as to whether frozen vegetables are better or worse than fresh ones. Generally, reports show that frozen vegetables are as nutritionally beneficial when compared to fresh ones.

A 1997 study performed by the University of Illinois, 2007 study performed by University of California - Davis and a 2003 Austrian study support that canned or frozen produce has no substantial nutritional difference not attributable to the presence of added salt, syrup or other flavouring, and in fact suggest that canned or frozen produce is nutritionally superior because of the very rapid deterioration of nutrients in fresh produce.

An advantage that frozen vegetables have over canned is that many brands contain little or no added salt because the freezing process by itself is able to stop bacterial growth. However, many canned vegetable brands with little or no sodium have become available and many frozen brands do have salt added for more flavour.

However, there may be some risk in eating poorly cooked frozen vegetables. For example, a 2007 Australian study found that frozen vegetables may contain a bacterium called Mycobacterium avium subspecies paratuberculosis (MAP) which is resistant to extreme cold and hot temperatures.

The history of frozen fruits can date back to the Liao Dynasty of China, with the "frozen" pear being a classic delicacy eaten by the Khitan tribes in the Northeastern region of China.

Canned Vegetables

Canned vegetables are a convenient and effective way of preserving vegetables for consumption when they are not readily available. Canned vegetables are full of essential nutrients, and in some cases the nutrients are more readily digestible than in the fresh equivalent. Most vegetables have naturally relatively high pH and thus fall into the low-acid group of foods. They therefore require a full sterilisation treatment, unless they are acidified. The kinds of microorganisms that can be found associated with most vegetables can cause food poisoning or thermophilic spoilage, and therefore it is very important that the sterilisation processes are well designed and carefully controlled.

Canned foods have a long shelf life, but that shouldn't mean that you keep them for several years before using them. High-acid canned foods such as juices, tomatoes, fruits and pickles will store well for 12 to 18 months. Whereas low-acid canned foods such as meat products and vegetables will store well for 2-4 years.However, there may be some changes in quality, such as a change in colour and texture.

A dry place with a moderately cool temperature is best to store canned foods. Make sure that you don't store them in a warm place near the oven or in direct sunlight. They should be kept dry and cool to prevent the can itself from rusting, which may cause leaks and eventually spoil the food inside.

It is not a recommended practice, but if it is necessary to heat canned food in the container, the top lid must be removed to prevent pressure build-up. The opened can may be covered loosely with a piece of aluminium foil, then placed in hot water and simmered. Never put a can in the microwave.

Look closely at all cans before opening them. A bulging lid, or a dented or leaking can is a sign of spoilage. When you open it, look for other signs, such as spurting liquid, an 'off' odour or mould. Don't taste or use canned foods that show any sign of spoilage. Throw them away immediately; they are not at all for consumption.

Once a can is opened, it becomes perishable and thus, should be either cooked properly or refrigerated. To preserve its flavour, opened canned foods should be transferred to plastic or glass containers before refrigerating. Leftovers should be consumed after 3 to 4 days of refrigeration.

Methods

The original fragile and heavy glass containers presented challenges for transportation, and glass jars were largely replaced in commercial canneries with cylindrical tin can or wrought-iron canisters (later shortened to "cans") following the work of Peter Durand. Cans are cheaper and quicker to make, and much less fragile than glass jars. Glass jars have remained popular for some high-value products and in home canning. Can openers were not invented for another thirty years — at first, soldiers had to cut the cans open with bayonets or smash them open with rocks. Today, tin-coated steel is the material most commonly used. Laminate vacuum pouches are also used for canning, such as used in MREs and Capri Sun drinks.

To prevent the food from being spoiled before and during containment, a number of methods are used: pasteurisation, boiling (and other applications of high temperature over a period of time), refrigeration, freezing, drying, vacuum treatment, antimicrobial agents that are natural to the recipe of the foods being preserved, a sufficient dose of ionizing radiation, submersion in a strong saline solution, acid, base, osmotically extreme (for example very sugary) or other microbially-challenging environments.

Other than sterilization, no method is perfectly dependable as a preservative. For example, the microorganism *Clostridium botulinum* (which causes botulism) can only be eliminated at temperatures above the boiling point of water.

From a public safety point of view, foods with low acidity (a pH more than 4.6) need sterilization under high temperature (116–130°C). To achieve temperatures above the boiling point requires the use of a pressure canner. Foods that must be pressure canned include most vegetables, meat, seafood, poultry, and dairy products. The only foods that may be safely canned in an ordinary boiling water bath are highly acidic ones with a pH below 4.6, such as fruits, pickled vegetables, or other foods to which acidic additives have been added.

Double Seams

Invented in 1888 by Max Ams, modern double seams provide an airtight seal to the tin can. This airtight nature is crucial to keeping micro-organisms out of the can and keeping its contents sealed inside. Thus, double seamed cans are also known as Sanitary Cans. Developed in 1900 in Europe, this sort of can was made of the traditional cylindrical body made with tin plate. The two ends (lids) were attached using what is now called a double seam. A can thus sealed is impervious to contamination by creating two tight continuous folds between the can's cylindrical body and the lids. This eliminated the need for solder and allowed improvements in manufacturing speed, reducing cost.

Double seaming uses rollers to shape the can, lid and the final double seam. To make a sanitary can and lid suitable for double seaming, manufacture begins with a sheet of coated tin plate. To create the can body, rectangles are cut and curled around a die, and welded together creating a cylinder with a side seam.

Rollers are then used to flare out one or both ends of the cylinder to create a quarter circle flange around the circumference. Precision is required to ensure that the welded sides are perfectly aligned, as any misalignment will cause inconsistent flange shape, compromising its integrity.

A circle is then cut from the sheet using a die cutter. The circle is shaped in a stamping press to create a downward countersink to fit snugly into the can body. The result can be compared to an upside down and very flat top hat. The outer edge is then curled down and around about 140 degrees using rollers to create the end curl.

The result is a steel tube with a flanged edge, and a countersunk steel disc with a curled edge. A rubber compound is put inside the curl.

Seaming

Opened can

The body and end are brought together in a seamer and held in place by the base plate and chuck, respectively. The base plate provides a sure footing for the can body during the seaming operation and the chuck fits snugly into the end (lid). The result is the countersink of the end sits inside the top of the can body just below the flange. The end curl protrudes slightly beyond the flange.

First Operation

Once brought together in the seamer, the seaming head presses a first operation roller against the end curl. The end curl is pressed against the flange curling it in toward the body and under the flange. The flange is also bent downward, and the end and body are now loosely joined together. The first operation roller is then retracted. At this point five thicknesses of steel exist in the seam. From the outside in they are:

- End
- Flange
- End Curl
- Body
- Countersink

Second Operation

The seaming head then engages the second operation roller against the partly formed seam. The second operation presses all five steel components together tightly to form the final seal. The five layers in the final seam are then called; a) End, b) Body Hook, c) Cover Hook, d) Body, e) Countersink. All sanitary cans require a filling medium within the seam because otherwise the metal-to-metal contact will not maintain a hermetic seal. In most cases, a rubberized compound is placed inside the end curl radius, forming the critical seal between the end and the body.

Probably the most important innovation since the introduction of double seams is the welded side seam. Prior to the welded side seam, the can body was folded and/or soldered together, leaving a relatively thick side seam. The thick side seam required that the side seam end juncture at the end curl to have more metal to curl around before closing in behind the Body Hook or flange, with a greater opportunity for error.

Seamer Setup and Quality Assurance

Many different parts during the seaming process are critical in ensuring that a can is airtight and vacuum sealed. The dangers of a can that is not hermetically sealed are contamination by foreign objects (bacteria or fungicide sprays), or that the can could leak or spoil.

One important part is the seamer setup. This process is usually performed by an experienced technician. Amongst the parts that need setup are seamer rolls and chucks which have to be set in their exact position (using a feeler gauge or a clearance gauge). The lifter pressure and position, roll and chuck designs, tooling wear, and bearing wear all contribute to a good double seam.

Incorrect setups can be non-intuitive. For example, due to the springback effect, a seam can appear loose, when in reality it was closed too tight and has opened up like a spring. For this reason, experienced operators and good seamer setup are critical to ensure that double seams are properly closed.

Quality control usually involves taking full cans from the line – one per seamer head, at least once or twice per shift, and performing a teardown operation (wrinkle/tightness), mechanical tests (external thickness, seamer length/height and countersink) as well as cutting the seam open with a twin blade saw and measuring with a double seam inspection system. The combination of these measurements will determine the seam's quality.

Use of a statistical process control (SPC) software in conjunction with a manual double-seam monitor, computerized double seam scanner, or even a fully automatic double seam inspection system makes the laborious process of double seam inspection faster and much more accurate. Statistically tracking the performance of each head or seaming station of the can seamer allows for better prediction of can seamer issues, and may be used to plan maintenance when convenient, rather than to simply react after bad or unsafe cans have been produced.

Pickled Vegetables

Most green vegetables may be preserved by pickling. Vegetables soften after 24 hours in a watery solution and begin a slow, mixed fermentation-putrefaction. The addition of salt suppresses undesirable microbial activity, creating a favourable environment for the desired fermentation. For example:

When the pickling process is applied to a cucumber, its fermentable carbohydrate reserve is turned into acid, its colour changes from bright green to olive or yellow-green, and its tissue becomes translucent. The salt concentration is maintained at 8 to 10 percent during the first week and is increased 1 percent a week thereafter until the solution reaches 16 percent. Under properly controlled conditions the salted, fermented cucumber, called salt stock, may be held for several years.

Salt stock is not a consumer commodity. It must be freshened and prepared into consumer items. In cucumbers this is accomplished by leaching the salt from the cured cucumber with warm water (43–54°C [110–130° F]) for 10 to 14 hours. This process is repeated at least twice, and, in the final wash, alum may be added to firm the tissue and turmeric to improve the colour.

Ways to Pickle Vegetables

Pickling Ingredients

There are so many different mixes and blends for pickling solutions that you could write an entire book on the subject (and some people have). You need some type of pickling vinegar; for best results try different types and depths of color. Don't be afraid to experiment. You'll also need pickling salt and some fresh garlic, so make sure you have these on hand before you get started.

Note that after pickling, the vinegar may change color. This is usually caused by the garlic you use, so try to use a smooth, blemish-free clove—and remove it before adding your vegetables.

Equipment Needed

There are several pieces of equipment you will need for pickling vegetables, depending on which method you choose. The quicker and easier method lets you store your pickled vegetables for around four months, the more complicated method using a pressure canner will allow longer storage.

For both methods, you'll need a large pan for boiling the pickling solution, a heat-proof jug and several sterilized jars with matching lids. You can sterilize these in a dishwasher on the shortest cycle (don't wash anything else with them, though) or in a medium-heated oven for 20 minutes or so, with the jars laid on their sides on a parchment paper-covered baking tray.

If you don't plan on refrigerating your pickled vegetables, you'll need to use a more sterile method which involves a pressure canner. These cost around $35 dollars and ensure the contents of your pickle jars remain safe and edible over a longer period of time.

Best Vegetables for Pickling

There are many vegetables suited for pickling. Zucchini and cucumbers are both popular choices—as are peppers, onions, carrots, cabbage. In fact, you can pickle pretty much any vegetable.

Wash, seed, and chop your vegetables as required. Then, parboil for about two minutes. Once the vegetables are removed from heat, add them to the jars for pickling.

With your vegetables prepared, boil the pickling solution. Meanwhile, remove your jars from the oven and carefully (using oven mitts) stand them up. Transfer the jars to a solid surface and begin to add your vegetables, leaving a 1/2 inch gap at the top of the jar. Pour in the pickling solution, again leaving a 1/2 inch gap at the top.

Place the lids on the jars, but wait for them to cool before tightening. Then, put the jars in the back of the fridge. With this method, your pickled vegetables will stay tasty for about four months. Of course, if you don't think you'll be able to eat them all within that time, you can still share with the neighbors.

Fermented Vegetables

Fermented vegetables begin with lacto-fermentation, a method of food preservation that also enhances the nutrient content of the food. The action of the bacteria makes the minerals in cultured foods more readily available to the body. The bacteria also produce vitamins and enzymes that are beneficial for digestion.

Almost any vegetable can be fermented, and fermenting farm-fresh produce is a great way to provide good nutrition year-round! Ferment one vegetable alone or create mix of many different kinds, along with herbs and spices, for a great variety of cultured foods.

Ways to Ferment Vegetables

1. Choose your fermentation equipment:

    While fermenting vegetables does not require a lot of specialized equipment, using the appropriate equipment can make all the difference when getting started. From a good chopping knife to the right fermentation vessel, you'll want to pick equipment to fit your needs. When choosing your fermentation equipment and supplies, consider your options carefully.

2. Prepare the vegetables for fermenting:

    There are several ways to prepare the vegetables for fermenting: grating, shredding, chopping, slicing, or leaving whole. How you choose to prepare your vegetables is a personal choice, though some vegetables are better suited for leaving whole, while others ferment better when shredded or grated.

3. Decide if you will use salt, whey, or a starter culture:

    A fermented food recipe may call specifically for salt, salt and whey, or a starter culture. The method chosen can vary, depending on personal taste, special dietary requirements, and even the vegetables used.

4. Use water to prepare the brine:

    Water used for preparing brine or starter culture should be as free from contaminants as possible, for the best-tasting fermented vegetables.

5. Weigh the vegetables down under the brine:

   Once the vegetables have been prepared and placed in the chosen fermentation vessel, weigh the vegetables down under the brine, keeping them in an anaerobic environment during the fermentation period.

6. Move the fermented vegetables to cold storage:

   Once the vegetables are finished culturing, it's time to move them to cold storage. When new to fermenting, it may be difficult to know exactly when to consider the vegetables finished.

7. Troubleshooting:

   As with any culturing process, each batch of fermented vegetables can turn out differently.

Eating Fermented Vegetables

Properly made fermented vegetables contain very high levels of probiotics, much higher than the best probiotic supplements. That makes them ideal for optimizing your gut flora and promoting the colonization and growth of beneficial bacteria.

The probiotics from fermented vegetables offer many potential benefits, including:

- Breaking down and eliminating toxins and wastes from your body
- Helping your body produce B vitamins and vitamin K_2
- Assisting in the absorption of minerals
- Helping you maintain your ideal weight
- Helping fend off disease and promoting wellness
- Improving your mood and promoting mental health.

References

- Clive Maier; Theresa Calafut (2008). Polypropylene: The Definitive User's Guide and Databook. Elsevier Science. ISBN 978-0-08-095041-9. Archived from the original on 4 December 2017
- Best-food-preservation-methods-for-fresh-vegetables-1413: homepreservingbible.com, Retrieved 14 July 2018
- Sobel J (October 2005). "Botulism". Clin. Infect. Dis. 41 (8): 1167–73. doi:10.1086/444507. PMID 16163636
- All-you-need-to-know-about-canned-foods-436287: nestle-family.com, Retrieved 29 March 2018
- "American Heart Association 2010 Dietary Guidelines" (PDF). 2010 Dietary Guidelines. American Heart Association. 23 January 2009. Archived from the original (PDF) on 24 January 2011. Retrieved 16 May 2010
- How-to-pickle-vegetables: tablespoon.com, Retrieved 21 May 2018
- Danesi, F.; Bordoni, A. (2008). "Effect of Home Freezing and Italian Style of Cooking on Antioxidant Activity of Edible Vegetables". Journal of Food Science. 73 (6): H109–12. doi:10.1111/j.1750-3841.2008.00826.x. PMID 19241586
- Fermented-vegetables-recipes: probiotics.mercola.com, Retrieved 16 June 2018

- Rickman (2007). "Nutritional comparison of fresh, frozen, and canned fruits and vegetables". doi:10.1002/jsfa.2824. Archived from the original on 11 April 2013. Retrieved 19 February 2012

- Canned-vegetables, food-science: sciencedirect.com, Retrieved 18 May 2018

- "Dietary Guidelines focus on sodium intake, sugary drinks, dairy alternatives". Food Navigator-usa.com. Decision News Media. 27 April 2010. Archived from the original on 6 May 2010. Retrieved 16 May 2010

Permissions

Index

www.ingramcontent.com/pod-product-compliance
Lightning Source LLC
Chambersburg PA
CBHW061241190326

41458CB00011B/3542